Tourism in Natural and Agricultural Ecosystems in the Eighteenth and Nineteenth Centuries

This book analyzes the roots of one of the main human activities that can be developed in natural and agricultural ecosystems: tourism.

Attention to natural and agricultural ecosystems and their conservation has intensified in recent decades, responding to increasing social sensitivity to the environment, as also witnessed by Agenda 2030. The book explores the development of tourism in natural and agricultural ecosystems in the eighteenth and nineteenth centuries, when some of its essential features derived from the practices of exploration, scientific study, business, healing practices, and also a desire for personal growth. This research is intended to open up international scholarly debate and discussion and draw in contributions from all disciplines and geographical areas. In addition, it intends to add an important piece to the mosaic of international literature that has rarely considered the origins of nature and rural tourism in an array of practices not always embodying a stated intent of recreation. This book is based on handwritten documents and travelogues circulating during the period in question. Most of the travel experiences analyzed regard men and women of European descent, but their travels were global, with ecosystems considered on all populated continents.

This volume is essential reading for students and scholars alike interested in tourism history and the history of science and travel.

Martino Lorenzo Fagnani is a postdoctoral researcher in early modern history at the University of Pavia. His main research interests are the history of science and the history of travel.

Luciano Maffi is a lecturer in economic and global history at the University of Parma, Italy. He also teaches economic history and tourism history at the Catholic University of Milan.

Routledge Studies in Modern History

For more information about this series, please visit: www.routledge.com/ Routledge-Research-in-Modern-History/book-series/MODHIST

Tourism in Natural and Agricultural Ecosystems in the Eighteenth and Nineteenth Centuries

Martino Lorenzo Fagnani and Luciano Maffi

Routledge
Taylor & Francis Group

NEW YORK AND LONDON

First published 2024
by Routledge
605 Third Avenue, New York, NY 10158

and by Routledge
4 Park Square, Milton Park, Abingdon, Oxon, OX14 4RN

*Routledge is an imprint of the Taylor & Francis Group, an informa
business*

ISBN: 978-1-032-13704-9 (hbk)
ISBN: 978-1-032-13706-3 (pbk)
ISBN: 978-1-003-23051-9 (ebk)

DOI: 10.4324/9781003230519

Typeset in Times
by SPi Technologies India Pvt Ltd (Straive)

Contents

Figures

Tables

Acknowledgments

Our book is the fruit of many years of work and would never have seen the light of day if we had not had the opportunity to discuss the results of our research in numerous contexts, receiving advice on how to improve them and suggestions for supplementing them with the works of other experts on these subjects and in associated branches of historical science.

We have presented our studies of tourism in natural and agricultural ecosystems at the European Society for Environmental History Conference organized in Bristol in July 2022 and at the European Social Science History Conference organized in Gothenburg in April 2023. We have also discussed our findings in numerous workshops on the history of tourism organized by the Milan and Brescia campuses of the Catholic University of the Sacred Heart.

During these years of study we have had the privilege of immersing ourselves in the history of travel, tourism, and science; economic history; and cultural history with many highly competent people to whom we are deeply indebted. We would like to thank Donatella Strangio, Manuela Mosca, Annunziata Berrino, Carlos Larrinaga, Rafael Vallejo Pousada, Patrizia Battilani, Petra Kavrečič, Sergio Onger, Andrea Zanini, Paul Warde, Emma Spary, Staffan Müller-Wille, Laurent Brassart, Lavinia Maddaluno, Monica Azzolini, Anna Pellegrino, Marina Romani, Agnese Visconti, Manuel Vaquero Piñeiro, Paolo Tedeschi, Luca Mocarelli, Giulio Ongaro, Emanuele Camillo Colombo, Andrea Maria Locatelli, Claudio Lorenzini, Omar Mazzotti, Dario Dell'Osa, Stefano Levati, Paola Bianchi, Giulia Giannini, Carlo Capra, Stefano d'Atri, Marco Armiero, Julia Adeney Thomas, Óscar Recio Morales, Juan Pan-Montojo, Ofelia Rey Castelao, Jean-Pierre Williot, Allen James Grieco, Ilaria Berti, Giacomo Bonan, Marianna Astore, Matti La Mela, Marco Emanuele Omes, and Carlo Baderna.

We would also like to thank our departmental colleagues. Specifically, in the Department of Humanities of the University of Pavia: Mario Rizzo, Davide Maffi, Matteo Di Tullio, Anna Rosa Candura, Alessandra Ferraresi, Giovanni Vigo, and Francesco Torchiani. In the Department of Economics and Management at the University of Parma: Gian Luca Podestà, Giovanni Ceccarelli, Stefano Magagnoli, Alberto Grandi, and Claudio Bargelli.

Our work as lecturers and researchers has allowed us to collaborate with scholars at the Catholic University of Milan and Brescia and at Bocconi University. At the former, we would like especially to thank Giovanni Gregorini, Maria Paola Pasini, Riccardo Semeraro, and Mario Taccolini. And our gratitude also goes to Guido Alfani, Anna Maria Monti, Mattia Fochesato, Marina Nicoli, Veronica Binda, Giorgio Bigatti, and Valeria Giacomin at the Bocconi University.

We are also indebted to the staff of the historical archives and institutes who helped us locate and study the handwritten documents and the old printed texts: the *Cancelliere* Rita Pezzola and the archivists Corrado Vailati and Maurizio Ghislandi of the Istituto Lombardo Accademia di Scienze e Lettere of Milan; Flora Bonalumi of the Biblioteca Nazionale Braidense; Director Marzia Dina Pontone, Antonella Campagna, and Elettra De Lorenzo of the Biblioteca Universitaria of Pavia; Cesare Repossi and Xenio Toscani of the Archivio Storico Diocesano of Pavia; Florence Tessier of the Botanical Library of the Muséum National d'Histoire Naturelle of Paris; Federica Bianchi, Chiara Pisani, and Roberta Benedusi of the Biblioteca Comunale Teresiana of Mantua; Esther García Guillén, Irene Fernández de Tejada de Garay, Abel Blanco Asenjo, and Gloria Perez de Rada Cavanilles of the Archive of the Real Jardín Botánico of Madrid.

We also thank Robert Burns for his meticulous linguistic revision of the text, Carlo Fagnani for his patient and excellent support with images, and Max Novick and Louise Ingham for their guidance as representatives of the publishing house.

Martino Lorenzo Fagnani
Luciano Maffi
Milan, April 21, 2023

And on a more personal note:

I would like to express my deepest gratitude to my parents for their constant support and to a number of people for the unstinting friendship you have always shown me: Francesca Martini, Carla Rubio, Franco Pavesi, Raniero Covi, and Giulia Rossi.

Martino

I would like to thank my parents, as well as Sara Cavanna, Alberto Anelli, Francesco Debiaggi, and Matteo Pirola, for the great benefit and honor of your enduring affection and patience.

Luciano

Abbreviations

ADG	Archivio Doria, Dipartimento di Economia dell'Università. Genoa
AIL	Archivio dell'Istituto Lombardo Accademia di Scienze e Lettere. Milan
ARJB	Archivo del Real Jardín Botánico. Madrid
ASDPv	Archivio Storico Diocesano. Pavia
BCMHN	Bibliothèque Centrale du Muséum National d'Histoire Naturelle. Paris
BCT	Biblioteca Comunale Teresiana. Mantua
BNB	Biblioteca Nazionale Braidense. Milan
BUPv	Biblioteca Universitaria. Pavia

Note

Except as otherwise indicated, citations from primary handwritten or printed sources originally written in Italian, French, or Spanish have been translated into English by the authors with the support of a specialized native English speaker.

Introduction

Today, attention to environmental issues, the protection of ecosystems, and eco-friendly activities—from agriculture to recreation—are addressed with increasing urgency by international organizations, countries, and public opinion. Not surprisingly, this kind of care in the approach to nature and the sustainable development of human activities is one of the fundamental objectives of the 2030 Agenda, adopted by all United Nations member states in 2015. Tourism is one of the activities with the greatest impact on the environment, cultures, and the global economy. In particular, among the objectives of the 2030 Agenda, there is also the global promotion of sustainable tourism that encourages local employment, culture, and production, as well as the development of tools to monitor the impact of this kind of tourism.[1]

From a historiographical point of view, tourism as a general phenomenon has acquired considerable space in recent decades, involving scholars from different geographical areas, promoting interdisciplinary approaches, and favoring the use of comparative study models. However, while tourism as a sustainable activity and its links with the environment, natural resources, and agriculture are studied by various disciplines, the attention devoted to it is still very limited in historical research and over a long-term perspective.[2]

We want to fill this gap, offer ideas for future debates, and provide a contribution to historical tourism studies, analyzing a series of sources over a period of two centuries.

The modern-day tourism categories of principal interest in our book are nature tourism and rural tourism. In its essential components, nature tourism is based on the natural attractions of an area; responsible, with a minimized impact on the local ecosystem; and beneficial to local communities living in harmony with that ecosystem. An outgrowth of nature tourism is ecotourism, which embraces the previously mentioned components but incorporates an element of sensitization and awareness-raising regarding the fragility of the destinations. Recent scientific literature contains many case studies on the impact of nature-based tourism on a broad range of

DOI: 10.4324/9781003230519-1

ecosystems and human activities, critically assessing its compatibility with principles of environmental sustainability.[3]

As for rural tourism, it shares many of the same dynamics as nature tourism but—as suggested by the name itself—refers to agricultural environments, products, and lifestyles. Rural tourism offers a way to counterbalance the seasonal productivity variations in the agricultural economy. The anthropic component is thus a main element of rural tourism, whereas in nature tourism, it is incidental if not completely absent and certainly not a distinctive feature.[4]

As for their origins, the advent of nature tourism is conventionally considered to date to the establishment of the major National Parks in North America, most notably Yellowstone in 1872, Sequoia and Yosemite in 1890, and Mount Rainier in 1899.[5] Rural tourism, by comparison, began to be codified much later. For example, after World War II, tourism began to be seen as an increasingly important economic resource in many rural areas of Europe. The first detailed legislation in individual EU member states began to emerge near the end of the twentieth century.[6]

The question we explore in this book is whether we can find the roots of nature tourism and rural tourism in an earlier period, tracing them to the eighteenth century. During these centuries, a number of scientific journeys were organized both in the European periphery and on other continents with the goal of expanding knowledge in the fields of botany, zoology, geology, and mineralogy, that is all branches generally of the natural sciences and geography in that period. At the same time, journeys not ascribable to scientific objectives were undertaken in natural settings. Here the travelers were writers, philosophers, memorialists, and artists, who set forth mainly for purposes of personal growth and sometimes for other reasons, often taking an interest in nature, collecting notes, and publishing accounts that enjoyed a fair amount of success. Furthermore, with the globalization ushered in by the Ocean Age (1500–1800), an increasing number of technicians and consultants were sent to work in distant lands, often for the colonies of their home country, sometimes for foreign powers. Frequently they were engaged in the development of natural or agricultural resources.[7]

Personal growth and recreation were the stated motives in only a handful of the journeys addressed in this book. In these cases, mainly in the category of humanist travelers, these desires were manifested from the early stages of planning. However, in most cases we discuss herein—scientific exploration, technical personnel assessing natural resources for the government, and missionary work—these benefits were an incidental outcome. Indeed, we may define this latter category as 'incidental tourism', that is, opportunities for the personal fulfillment and recreation we associate with tourism today that occur as ancillary effects of travel undertaken for other some central purpose, such as work, scientific research, professional training, medical care, or religious pursuits. In most cases, the desire for diversion and edification were secondary objectives at best.[8]

Among the cases we have studied for this work are those where the protagonist remained abroad for a long period, establishing a base in a foreign land that could be used as the starting point for other travels. Attention to natural aspects, ecosystems, and the human impact on natural equilibria is evident in accounts published as scientific treatises or memoirs and in correspondence or private diaries. While we would be amiss in speaking of ecological sensitivity, an awareness of the qualities of life outside of the urban dimension and the experience of new emotions in either a natural or a rural setting were clearly communicated in the sources we have examined.[9]

There is an extensive historiography of travel analyzing its development within the context of natural sciences, education, economy, politics, religion, and other spheres. What is missing, however, is focused research examining the experience of pristine or anthropized nature as a precursory element for certain types of modern-day tourism: not only nature-based tourism and rural tourism first and foremost but also the various subcategories of experiential tourism. Examples include pescatourism, where the traveler gains an experience of the life of fishing communities while providing them with an alternative source of income in various parts of the planet,[10] and extreme tourism, thrilling experiences in particularly wild and unprotected settings.[11]

We also compare our case studies to certain aspects of contemporary concepts of nature and ecology. Similarities may be found between some of the travel experiences in natural and rural ecosystems that we examine in our book and tourism as we understand it today. Issues relating to the relationship between human society and local fauna are one example. We review the difficulties associated with the coexistence of humans and crocodiles in the fluvial and coastal areas of Sri Lanka: while the latter are potentially dangerous to humans, they are also a characteristic aspect of Sri Lankan nature. These elements are noted both in historical and contemporary accounts, as we will discuss.

Chapter 1 proposes a categorization of travelers into three groups: scientists, humanists, and working travelers. It analyzes the different approaches to ecosystems influenced by their respective backgrounds and the objectives of their journey. The category of scientists includes naturalists and agriculturalists visiting different parts of the world in the eighteenth and nineteenth centuries. In many cases, they played an important role in scientific expeditions both in wild areas of Europe and other continents, which still contained vast uncharted tracts at the time. To the humanist category belong writers, memorialists, artists, and intellectuals mainly interested in describing the sensations triggered in them by the ecosystems they encountered and the cultural aspects between these ecosystems and the local populations, at times with the stated objective of presenting them to a broad audience in their published works. The third category comprises all those traveling in the employment of others on a practical mission that required their specific expertise, including business and missionary travel.

In spite of the difference in backgrounds and objectives, all these travelers shared a fascination for the ecosystems they visited, at times for their components and at times for the sentiments aroused by the general experience. Our analysis in Chapter 1 was guided by a search for similarities and differences in an attempt to identify precursory elements of nature tourism and rural tourism in travels that were often dictated by reasons of work, scientific research, or study.

Chapter 2 distinguishes between the types of environments visited in travels that may be considered precursors to modern nature tourism and rural tourism: agricultural, Alpine, glacial, desert, marine, and 'colonial'. We also added a section focusing on urban parks and gardens, which became increasingly common and varied. All these types of environments aroused strong interest in those traveling for both professional and recreational reasons. In this chapter, we focus our analysis on the aspects of these environments that were most striking to the travelers, regardless of the reason for their travel. We also consider their interaction with the local flora and fauna and any humans in the area, as well as how they dealt with obstacles to their journey imposed by the characteristics of each particular environment.

Chapter 3 is dedicated to how travels to natural and agricultural ecosystems were organized. Here we isolate three specific components on which to focus our analysis: transportation, food and lodging, and contact with locals. For much of the two centuries we study, travel was relatively difficult and dangerous compared to its modern counterpart due to undeveloped infrastructure that was difficult to maintain and make safe, complicated systems for certifying one's identity and obtaining transit permits that were not always within everyone's reach, the difficulty in preserving food, and the need to protect oneself from diseases. These issues, of course, were compounded in distant lands with extensive wilderness areas.

The case studies in Chapter 3 allow us to assess the scope of these difficulties, the strategies adopted to overcome them, and especially the opportunities that these difficulties—paradoxically—presented for experiencing the natural ecosystems. It is also interesting to note over the long term how particularly challenging and adventuresome routes have now become the venues for extreme tourism and experiential tourism in 'pristine' natural settings. In other cases, travel management challenges arose in more anthropized contexts, obliging travelers to have more direct contact with the local population and production systems. This allowed a more direct and original experience of local foods and customs, which are sought-after characteristics in rural tourism experiences.

Chapter 4 addresses the role of written materials—both handwritten and printed—in facilitating and supporting travels in natural and rural ecosystems in the two centuries in question. These materials include travel guides, travel narratives, maps, and astronomical charts. We also emphasize the large bibliographies assembled over the years by travelers, often including scientific texts on the flora, fauna, and minerals of distant lands, most of them written by authorities on the subjects who thus provided a rather in-depth description of the ecosystems that would be visited. There were also historical and anthropological works outlining the relations between the local human populations and their environments. We assess how the circulation of these written materials influenced travelers' views of the natural and rural settings they encountered.

Our research method is predominantly qualitative. Our sources include handwritten documents in Italian, Spanish, and French, many unpublished. We also consulted travel notebooks, personal correspondence, notes for scientific works, and lists and notes written in preparation for a given journey. This documentation is interwoven with the numerous travel narratives published during the eighteenth and nineteenth centuries that we use as our primary printed sources because they provide direct testimony on regions all over the world visited in that period.

Among recent studies that have addressed the history of travel in the eighteenth and nineteenth centuries from a range of perspectives, we mention *The Cambridge History of Travel Writing*, edited by Nandini Das and Tim Youngs, and works written by Sarah Goldsmith and Judith Johnston.[12] Regarding the specific case of travel as a fundamental element of research in the natural sciences, strategies for improving agriculture, and scientific training (and the role of this type of travel in producing travel narratives and literary works), the recent collections of essays edited by Simona Boscani Leoni, Sarah Baumgartner, and Meike Knittel; by Frazer MacDonald and Charles Withers; and by Judy Hayden, as well as the books written by Nina Gerassi-Navarro and by Paul Smethurst, helped us build a solid methodological framework.[13] It should be noted that this bibliography addresses travel for purposes of training, exploration, and mapping of natural and agricultural resources. Indeed, the only historical meaning that comes close to the modern concept of tourism is that of the Grand Tour.

Peter Shackleford has also recently written an updated history of the World Tourism Organization, tracing the development of this specialized agency of the United Nations from its inception in 1925 to the modern day. While not encompassing the historical period we examine in this book—when travel in natural environments sowed the seeds of the modern concept of ecotourism, nature-based tourism, rural tourism, and forms of experiential tourism where interaction with nature is key—this work is nevertheless a valuable source in the methodological and interpretive framework applied in our study.[14]

Contributing to solidifying this framework as regards environmental history and the development of the concept of sustainability are the studies collected in the *Routledge Handbook of the History of Sustainability*, edited by Jeremy Caradonna; *The Oxford Handbook of Environmental History*, edited by Andrew Isenberg; *Views from the South*, edited by Marco Armiero; as well as *Les révoltes du ciel* written by Jean-Baptiste Fressoz and Fabien Locher; *The Environment: A History of the Idea* by Paul Warde, Libby Robin, and Sverker Sörlin; *The Invention of Sustainability* by Warde; *Nature and Power: A Global History of the Environment* by Joachim Radkau; and *The Unending Frontier* by John F. Richards.[15]

There are also studies directly addressing nature tourism, rural tourism, and other types mentioned earlier, such as scientific tourism, extreme tourism, and experiential tourism. However, they have mainly been studied in sociological, anthropological, and economic terms.[16] A wide-ranging, well-organized, and purely historiographical perspective is still lacking. We hope to make a significant contribution to this branch of research.

Notes

1 See the page *Tourism in the 2030 Agenda* in the UNWTO website, https://www. unwto.org/tourism-in-2030-agenda, accessed on December 22, 2022.
2 Julia N. Albrecht (ed), *Managing Visitor Experiences in Nature-Based Tourism* (Wallingford and Boston, 2021); Peter Fredman and Jan Vidar Haukeland (eds), *Nordic Perspectives on Nature-Based Tourism: From Place-Based Resources to Value-Added Experiences* (Cheltenham UK and Northampton MA, 2021); David A. Fennell, *Ecotourism* (Abingdon and New York, 2020, 5th edition); Andrew Holden and David Fennell (eds), *The Routledge Handbook of Tourism and Environment* (Abingdon and New York, 2013); Clem Tisdell and Clevo Wilson, *Nature-Based Tourism and Conservation: New Economic Insights and Case Studies* (Cheltenham UK and Northampton MA, 2012); Caroline Kuenzi and Jeff McNeely, *Nature-based tourism*, in Ortwin Renn and Katherine D. Walker (eds), *Global Risk Governance: Concept and Practice Using the IRGC Framework* (Dordrecht, 2008), pp. 155–178; C. Michael Hall and Stephen Boyd (eds), *Nature-Based Tourism in Peripheral Areas: Development or Disaster?* (Clevedon, Buffalo and Toronto, 2005).
3 For instance, the cases of Sweden and Finland are in-depth analyzed in Juulia Räikkönen, Miia Grénman, Henna Rouhiainen, Antti Honkanen, Ilari E Sääksjärvi, 'Conceptualizing Nature-Based Science Tourism: A Case Study of Seili Island, Finland', *Journal of Sustainable Tourism* (2021), https://doi.org/10. 1080/09669582.2021.1948553, accessed on December 22, 2022; Lusine Margaryan, 'Nature as a Commercial Setting: The Case of Nature-Based Tourism Providers in Sweden', *Current Issues in Tourism*, Volume 21, No. 16 (2018), pp. 1893–1911. For another example, this time related to Central American

countries, see Andrew P. Miller, *Sustainable Ecotourism in Central America: Comparative Advantage in a Globalized World* (Lanham and London, 2016).

4 Mary Cawely and Desmond A. Gillmor, 'Integrated Rural Tourism: Concepts and Practice', *Annals of Tourism Research*, Volume 35, No. 2 (2008), pp. 316–337; Brian Garrod, Roz Wornell, Ray Youell, 'Re-Conceptualising Rural Resources as Countryside Capital: The Case of Rural Tourism', *Journal of Rural Studies*, Volume 22, No. 1 (2006), pp. 117–128; Franz Hackl, Martin Halla, Gerald J. Pruckner, 'Local Compensation Payments for Agri-Environmental Externalities: A Panel Data Analysis of Bargaining Outcomes', *European Review of Agricultural Economics*, Volume 34, No. 3 (2007), pp. 295–320; Derek R. Hall, Lesley Roberts, Morag Mitchell (eds), *New Directions in Rural Tourism* (Hants, 2005, 2nd edition); Gunjan Saxena, Gordon Clark, Tove Oliver, Brian Ilbery, 'Conceptualizing Integrated Rural Tourism', *Tourism Geographies*, Volume 9, No. 4 (2007), pp. 347–370; David Weaver, 'The Distinctive Dynamics of Exurban Tourism', *International Journal of Tourism Research*, Volume 7 (2005), pp. 23–33.

5 Lynn Ross-Bryant, *Pilgrimage to the National Parks: Religion and Nature in the United States* (Abingdon and New York, 2013); Richard Grusin, *Culture, Technology, and the Creation of America's National Parks* (Cambridge and New York, 2004).

6 Bill Slee, *Rural tourism and recreation in the EU*, in Sophia M. Davidova, Kenneth J. Thomson, Ashok K. Mishra (eds), *Rural Policies and Employment: TransAtlantic Experiences* (Singapore, 2019), pp. 333–350; Ana María Campón-Cerro, José Manuel Hernández-Mogollón, Helena María Baptista Alves, Elide Di-Clemente, *The tourist in rural destinations: an experiential approach based on relationships with local people and surroundings*, in Elisabeth Kastenholz, Maria João Carneiro, Celeste Eusébio, Elisabete Figueiredo (eds), *Meeting Challenges for Rural Tourism through Co-Creation of Sustainable Tourist Experiences* (Newcastle upon Tyne, 2016), pp. 103–131.

7 Sarah Goldsmith, *Masculinity and Danger on the Eighteenth-Century Grand Tour* (London, 2020); Jeffrey D. Sachs, *The Ages of Globalization: Geography, Technology, and Institutions* (New York, 2020); Ana Simões, Ana Carneiro, Maria Paula Diogo (eds), *Travels of Learning: A Geography of Science in Europe* (Dordrecht, Boston and London, 2003).

8 Giuseppe Rocca, *Dal prototurismo al turismo globale. Momenti, percorsi di ricerca, casi di studio* (Turin, 2013), pp. 370–371.

9 For the evolution of concepts such as 'ecology' and 'sustainability' in the early modern period, refer to Paul Warde, *The Invention of Sustainability: Nature and Destiny, c. 1500–1870* (Cambridge and New York, 2018); Paul Warde, *Ecology, Economy and State Formation in Early Modern Germany* (Cambridge and New York, 2006).

10 Pescatourism is to be distinguished from recreational fishing. Vahdet Ünal, Irmak Ertör, Pinar Ertör-Akyazi, Sezgin Tunca, *Making pescatourism just for small-scale fisheries: the case of Turkey and lessons for others*, in Svein Jentoft, Ratana Chuenpagdee, Alicia Bugeja Said, Moenieba Isaacs (eds), *Blue Justice: Small-Scale Fisheries in a Sustainable Ocean Economy* (Cham, 2022), pp. 315–333.

11 Steve Taylor, Peter Varley, Tony Johnston (eds), *Adventure Tourism: Meanings, Experience, and Learning* (Abingdon and New York, 2013); Godfrey Baldacchino (ed), *Extreme Tourism: Lessons from the World's Cold Water Islands* (Kidlington and Amsterdam, 2006).

12 Nandini Das and Tim Youngs (eds), *The Cambridge History of Travel Writing* (Cambridge and New York, 2019); Goldsmith, *Masculinity and Danger*; Judy Johnston, *Victorian Women and the Economies of Travel, Translation, and Culture, 1830–1870* (Abingdon and New York, 2016, 2nd edition).

13 Simona Boscani Leoni, Sarah Baumgartner, Meike Knittel (eds), *Connecting Territories: Exploring People and Nature, 1700–1850* (Leiden and Boston, 2022), Fraser MacDonald and Charles W. J. Withers (eds), *Geography, Technology and Instruments of Exploration* (London and New York, 2015); Judy A. Hayden (ed), *Travel Narratives, the New Science, and Literary Discourse, 1569–1750* (Farnham and Burlington VT, 2012). See also Nina Gerassi-Navarro, *Women, Travel, and Science in Nineteenth-Century Americas: The Politics of Observations* (Cham, 2017); Paul Smethurst, *Travel Writing and the Natural World, 1748-1840* (Basingstoke and New York, 2012).

14 Peter Shacklefold, *A History of the World Tourism Organization* (Bingley, 2020).

15 Jeremy L. Caradonna (ed), *Routledge Handbook of the History of Sustainability* (Abingdon and New York, 2018); Andrew C. Isenberg (ed), *The Oxford Handbook of Environmental History* (Oxford, 2014); Marco Armiero, *Views from the South: Environmental Stories from the Mediterranean World (19th–20th centuries)* (Naples, 2006); Jean-Baptiste Fressoz and Fabien Locher, *Les révoltes du ciel: une histoire du changement climatique XVᵉ-XXᵉ siècle* (Paris, 2020); Paul Warde, Libby Robin, Sverker Sörlin (eds), *The Environment: A History of the Idea* (Baltimore, 2018); Warde, *The Invention of Sustainability*; Joachim Radkau, *Nature and Power: A Global History of the Environment*, trans. Thomas Dunlap (New York, 2008); John F. Richards, *The Unending Frontier: An Environmental History of the Early Modern World* (Berkeley, Los Angeles, and London, 2003).

16 Joseph S. Chen and Nina K. Prebensen (eds), *Nature Tourism* (Abingdon and New York, 2017); Susan L. Slocum, Carol Kline, Andrew Holden (eds), *Scientific Tourism: Researchers and Travellers* (Abingdon and New York, 2015); Katia Laura Sidali, Achim Spiller, Birgit Schulze (eds), *Food, Agri-Culture and Tourism: Linking Local Gastronomy and Rural Tourism: Interdisciplinary Perspectives* (Berlin and Heidelberg, 2011).

1 Travelers

1.1 A Methodological Note

The starting point of our analysis of nature and rural tourism is the study of traveler profiles. Depending on the travelers' backgrounds and the reasons that led them to a given environment, their encounters with wild or domesticated nature occurred in different ways and had different effects on them.

Before introducing the criteria we use to categorize travelers by profile, we shall take a moment to discuss the question of duration. According to the definition provided by the World Tourism Organization (UNWTO), a tourist experience must last for more than twenty-four hours but not more than a year. However, given the state of transportation technology in the eighteenth and nineteenth centuries and the relatively large amount of time it took to travel from one part of the world to another, this criterion may not be fully applicable to that period.[1]

For example, would an experience like Mary Wollstonecraft Shelley's travels in Italy from 1818 to 1822, in the context of the Grand Tour, be considered tourism today according to the UNWTO's temporal criterion? In this specific case, it must be borne in mind that the writer moved between different Italian cities and towns, residing in some for many months. In other cases we examine, a person moved to another country, settled in a town for an extended period, and from there set out on a number of shorter travel experiences. Another example in this chapter is the Swedish naturalist Pehr Kalm's sojourn in North America from 1748 to 1751, during which he made a number of expeditions.[2]

To return to the proposed subdivision by traveler profile, this is one of many that can be adopted in an analysis of *ante litteram* forms of nature and rural tourism and science tourism to some extent. For example, a division has been suggested between naturalistic training trips, often intertwined with shorter or longer stays in important scientific centers to learn from preeminent scholars and major collections, and true research trips, the main

DOI: 10.4324/9781003230519-2

purpose being precisely to study the environment and natural resources of a territory by collecting plant, animal, and mineral samples. However, such a dichotomy cannot exclude overlaps.[3] A similar ambiguity may be found with other approaches. We believe that a subdivision based on traveler profiles allows for a more distinct separation and broader inclusion in analyzing the primitive forms of natural and rural tourism.

However, this subdivision must not be interpreted as a rigid grid: some case studies with a certain degree of hybridism will emerge, which we have decided to include more in one category than in another because their training and the objectives of their habit to travel they suggest that, but other characteristics of their profile also bring them closer to other categories.

We propose three macro-categories: scientists, humanists, and working travelers. This tripartite division based on training, sociocultural background, travel objectives, and reactions to natural or rural settings gained form and definition via our analysis of the chosen case studies.

1.2 Scientists

While the term *scientist* is partly anachronistic for the period considered in our analysis, we use it to conveniently label a category that includes naturalists and agriculturists, that is, experts who needed to travel in natural or rural, pristine or anthropic, settings because of their particular field of study. The goal was to collect plant, animal, and mineral samples; study natural phenomena; and collect information on the interaction of populations with the local flora, fauna, and environments. Sometimes the naturalists and agriculturists traveled of their own accord, to enrich their stock of information, gain experience, and perfect their knowledge. Socioeconomic status was an important factor here, since not everyone had the wherewithal and freedom to take such wide-ranging, prolonged, and expensive travels. In many cases, however, the journey of naturalists and agriculturists was funded by governments and organized by one or more scientific institutions. An exemplary case is that of the expeditions to Latin America and the Philippines sponsored by the Spanish Crown.[4]

The profiles that we take into consideration are those represented in Table 1.1, which indicates nationality and destinations.

The predominant scientific field was that of botany, to be understood both as a pure discipline and as applied to the study of agriculture. The interest in the plant kingdom alongside that for geology made this type of traveler particularly alert to details of scientific value. We have also noted a certain sensitivity to the beauty of the natural landscape among botanists, who thus tended to be more emotionally engaged with the surrounding environment.

Table 1.1 Travelers analyzed in Section 1.2. Our elaboration

Name	Life	Nationality	Travels analyzed in Chapter 1
Carl Linnaeus	(1707–1778)	Swedish	Northern Europe
Lazzaro Spallanzani	(1729–1799)	Italian	Southern Europe
Pehr Kalm	(1716–1779)	Swedish	North America
Anna Maria Walker Patton	(1778–1852)	British	Indian Ocean
Charles Darwin	(1809–1882)	British	South America and Pacific Ocean
Alberto Fortis	(1741–1803)	Italian	Southern Europe
Antonio José Cavanilles	(1745–1804)	Spanish	Southern Europe
Alexander von Humboldt	(1769–1859)	German	South America and Atlantic Ocean
Enrico Hillyer Giglioli	(1845–1909)	Italian	East Asia

We find various eighteenth- and nineteenth-century accounts of uncontaminated nature: more precisely, areas with limited anthropic presence, perhaps isolated structures used as a refuge during the journey. Carl Linnaeus himself offers us the first example of travel sharing certain characteristics with today's nature and science tourism. In particular, from May to October 1732, at the age of twenty-five, he undertook a scientific journey in northern Sweden, including some parts of modern-day Finland, to study its environment, plants, minerals, animals, and the customs of the local peoples. The expedition was financed by the Royal Society of Sciences of Uppsala, the city where the young Linnaeus was studying and teaching. He traveled partly on horseback and partly on foot, with the help of guides, interpreters, and servants.[5]

The travel report was translated from Swedish and Latin into English by James Edward Smith (1759–1828), founder in 1788 and first president of the Linnaean Society in London. Published in the British capital in 1811, *Lachesis Lapponica or a Tour in Lapland* contains various elements that demonstrate Linnaeus's emotional response to the surrounding environment, beyond his scientific observations. He describes the natural element sometimes as gloomy, oppressive, and indomitable, while at other times it makes the journey so pleasant that it suggests traits of an embryonic form of tourism.

The descriptions of the landscape in the countryside of Lycksele fall into the former category. Linnaeus reconstructs for the reader a land of barren sand dotted with scattered fir trees 'towering up to the clouds'. There were also patches of ground covered with moss and blueberry bushes. Birch trees and solid pines grew, but 'on the dry hills ... the finest timber was strewed around, felled by the force of the tempests, lying in all directions, so as to render the country in some places almost impenetrable. I seemed to have reached the residence of Pan himself'.[6] Equally gloomy is the description of the Scandinavian Mountains a few days later, where forests were absent and snow covered everything, leading him to comment that 'the verdure of summer seemed to shun this frozen region' and 'the delightful season of spring seemed an alien here'.[7]

Other passages had sweeter, more pleasant tones, with comments on the best times to visit some pleasant corners of the region. For example, there were hints of the beautiful forests of silver fir, spruce, and birch in the Piteå area, and no area of Sweden was better for summer travel (Linnaeus travelled in June).[8] In Lycksele, he had stopped to observe the abundance of bog-rosemary (with the suggestive scientific name *Andromeda polifolia*, chosen by Linnaeus himself) 'at this time in its highest beauty', which with the 'flesh-colour' of its corollas embellished the marshy grounds of that area. Beholding such natural wonder, Linnaeus could only conclude: 'Scarcely any painter's art can so happily imitate the beauty of a fine female complexion; still less could any artificial colour upon the face itself bear a comparison with this lovely blossom'.[9]

While at the top of the Piteå mountains, Linnaeus and his companions were able to admire green Norway at their feet like a 'garden in miniature'. Once they descended, they enjoyed the delightful sensation of 'the warmth and beauty of summer':

> The verdant herbage, the sweet-scented clover, the tall grass reaching up to my arms, the grateful flavour of the wild fruits, and the fine weather which welcomed me to the foot of the alps, seemed to refresh me in mind and body.[10]

Linnaeus's account also includes a wealth of attention to local culture and customs, an element that is fully part of the concept of nature tourism today. Alongside unspoiled nature, the taxonomist also expressed an interest in the cultural and historical background of the areas he visited, noting local activities and customs we would now term *sustainable*.[11]

For example, Linnaeus's descriptions of the eating habits of the inhabitants in the various areas visited were quite accurate. The inhabitants of the Torneå area—a city in modern-day Finland—prepared meals consisting of bread, fish (*strömming*), sometimes cheese and butter, cabbage, turnip, legumes, and sour milk, as well as some other simple ingredients.[12] The inhabitants of the Lycksele area had a much poorer diet, based on fresh or dried fish that was boiled, roasted, or made into soup ('I wondered … how these poor people could feed entirely on fish', Linnaeus noted astonished). From midsummer to autumn, they milked reindeer; in the more mountainous areas, they sometimes boiled this milk with sorrel to preserve it over the winter. In the cold season, they sometimes slaughtered reindeer, which thus ended up constituting another source of food, albeit scarce.[13]

More sanguine and reckless than Linnaeus in enjoying the natural beauty and taking an interest in local folklore was the Italian clergyman Lazzaro Spallanzani, professor of natural history at the University of Pavia (northern Italy) since 1769 and today considered one of the fathers of modern biology. He traveled extensively throughout his career to collect samples to add to the museum of natural history that he oversaw at the university.[14]

In the summer and autumn educational recess of 1788, almost sixty, he traveled to southern Italy with the main intent of collecting volcanic samples from areas such as the Campi Flegrei in what is now Campania and on Mount Etna and the Aeolian Islands in Sicily.[15] Spallanzani's interest in volcanoes was framed by the diatribe between Neptunists and Plutonists about the origin of rocks. The Neptunist geological theory, with its main exponent the German Abraham Gottlob Werner (1749–1817), considered the Earth's crust to be the product of the deposition of substances contained in the primordial ocean that had once covered the Earth. Plutonism, by contrast, under its main theorist, the Scotsman James Hutton (1726–1797), emphasized the key role of terrestrial heat, escaping via volcanic eruptions, in the formation of rocks; magmatic processes were therefore a key factor in the rock formation processes. Although Spallanzani may have been more adventurous and romantic than Linnaeus, his scientific ideas and objectives were broad and well grounded, including consideration of the Neptunism–Plutonism debate.[16]

The description he gave of the Neapolitans' relationship with Vesuvius is interesting. After arriving in the area on July 24 and observing signs of volcanic activity, Spallanzani immediately remarked that they did not awaken the curiosity of the Neapolitans, 'who, used to having it there before their eyes every day, are not in the habit of thinking about it, except when a major, destructive eruption occurs'.[17]

Relatively unaccustomed to Vesuvius and the spectacle of its activity, albeit not particularly pronounced at that time, Spallanzani was greatly impressed by what he saw: the constant smoke that enveloped the top of the mountain, sometimes carried by the wind as far as the island of Capri and, at night, flames and reddening fissures here and there, even without any audible detonations reaching Naples. Returning from Sicily in early November by sea, he was lucky enough to see a conspicuous flow of lava even before he reached Capri. He did not see Vesuvius 'at the worst of its fury' and was thus motivated to make a closer approach. He took up residence in a hermitage two miles from the mountain and watched it for hours in the clear, moonless night before going to bed. Getting up four hours before dawn, he approached the volcano on foot, admiring more closely 'the fiery jets', 'a cloud of flaming stone' and 'a great quantity of lively sparks' against a background of continuous detonations.[18]

As John Brewer and Sarah Goldsmith recently pointed out, Vesuvius was a must-see destination at least since the end of the seventeenth century for European travelers of all categories who came to Italy. The volcano was quite accessible and could be admired in all its natural charm. While Spallanzani's interest was primarily scientific, his encounter with Vesuvius was also an emotional experience. Indeed, at the beginning of the nineteenth century, the romantic writers included Vesuvius among the crucial sources of the sublime.[19]

At the beginning of September, Spallanzani had been in Sicily to visit Etna. After traversing its slopes, he spent the night with the guides in a cave eight miles from the summit. Three hours before dawn, they set out to reach it. It was still dark, but the guides' torches allowed him to glimpse the surrounding landscape: barren, stony, devoid of plants except for a few bushes here and there. Then the sun rose:

> I was four miles from the lips of the huge crater when I began to emerge from the shadows of the night and into the light of day. A faint glow in the east, a white light, an aurora rich in rosy color, the sun peeking over the horizon, initially turbid, flickering, and vaporous, then imperceptibly brighter, more resplendent; thus were the gradations of the nascent day, never seen elsewhere with such clarity and delight as on that lofty slope so close to the summit of Etna.[20]

In this case, the analytical description of a professional translates into a poetic description. He records what seems to be a primordial dawn in the heart of the Mediterranean from the millennia before Christ while purportedly on a scientific journey to bring samples back to the University of Pavia.

Observing the volcanic rocks he encountered with the expert eye of the naturalist, Spallanzani arrived two and a half miles from the summit of Etna, which was plumed with two high columns of smoke and surrounded by fumaroles. Here the path was interrupted by a solidified lava flow—it was from 1787, the guides assured him—which created a difficult and dangerous barrier. It was both jagged, full of spikes and holes, and slippery at points, so smooth that it resembled a sheet of ice. Although many months had passed, red-hot lava could still be glimpsed in some fissures with 'reddened recesses'. Spallanzani inserted his walking stick, which immediately caught fire. But not at all discouraged by the danger, he continued up to the top of the cone. Here he sat on the edge of the crater and triumphantly stayed there for two hours to rest and admire 'the interior of that theatrical volcano'.[21]

Pehr Kalm, a student of Linnaeus and professor of economics and director of the botanical garden at the Royal Academy of Turku (modern-day Finland, then in the Kingdom of Sweden), gives us our first example of extra-European travel. The Royal Academy of Sweden sponsored a journey by Kalm to North America with the main objective of obtaining red mulberry specimens for economic purposes. Kalm's expedition lasted from 1748 to 1751, during which he carefully studied the North American territory, customs, and flora and fauna. Kalm used the Swedish-Finnish town of Raccoon in New Jersey as his main base for some months. He visited vast territories of what is now the eastern United States and Canada.[22]

If his teacher Linnaeus, and Spallanzani in his journeys, focused above all on the botanical, geological, lithological, and orographic elements of the landscapes they visited, Kalm was also interested in unusual spectacles. For example, during the Atlantic crossing, he noted 'at night-time sparks of fire, as if floating on the water, especially where it was agitated, sometimes one single spark swam for the space of more than one minute on the ocean before it vanished'. He did not know it, but it was probably the bioluminescent dinoflagellate *Noctiluca scintillans* (Macartney) Kofoid & Swezy, 1921. Kalm did not merely observe and record his own experience of the phenomenon but also that of the crew: they had observed how the sparks 'appeared commonly during and after a storm from the north, and that often the sea was as if full of fire, and that some such shining sparks would likewise stick to the masts and sails'. While registering the wonder of the sailors before an ocean 'full of fire', the scientist in him observed the phenomenon from a more detached point of view, noting that the brightness sometimes 'looked rather like a phosphorescence of putrid wood'.[23]

In addition to his taste for mysterious and inexplicable details of nature, Kalm was also awed by landscapes. He had this to say about Niagara Falls:

> One cannot help being amazed and awed when standing above the cataract gazing down. It felt as if one were looking down from the highest church steeple. The falls that I had seen previously ... in North America, and ... in Sweden seemed but child's play in comparison. One could not gaze and contemplate without feelings of wonder and astonishment.[24]

Although the style of Kalm's accounts—published in Swedish in 1753–1761 and later translated into numerous languages—is more aseptic than Spallanzani's colorful narration of his travels, his words nevertheless clearly convey his fascination with the countries he visited and their natural environment and wildlife, expressed in an almost novelistic style.

Let us now turn to a case study from the mid-nineteenth century. The Scottish botanist Anna Maria Walker (née Patton) made an expedition in late January and early February 1833 to climb Adam's Peak (Sri Pada) in Ceylon (Sri Lanka) together with her husband, a colonel in the British Army, traveling as far as possible on horseback and then on foot. The trip lasted about two weeks and was part of the work to collect seeds and samples from the island to be sent to the botanical gardens in Edinburgh and Glasgow.[25]

Walker's account—published in 1835 by the *Companion of the Botanical Magazine* under the direction of William Jackson Hooker (1785–1865), professor of botany at the University of Glasgow—provides detailed descriptions of Sri Lankan flora, as to be expected given the nature of the journey and the magazine in which the account was published. However, alongside the scientific names and technical descriptions of the species encountered, she let herself be carried away by the 'most magnificent bamboos I ever saw', 'superb ferns', and 'a pretty white *Convolvulus*, which covers many of the bushes, and even trees, hanging in beautiful festoons, loaded with blossom, from branch to branch'.[26] Also impressive is the night view of February 3 when she reached the summit:

> At one A.M. we again ascended to the highest point—the mist was gone – the full moon shone bright—the scene was stupendous—the deep shadows making the hollows appear unfathomable, while the higher and more prominent features of the scene were illuminated by the mild and silvery lustre of a tropical moon, the most beautiful of all lights—of which none who have not seen it can form a conception....[27]

There were also hints of the pleasantness of the essential Western rites amid unspoiled nature, such as a luncheon at the foot of a group of 'magnificent' Ceylon ironwood trees (*Mesua ferrea* L.). However, as we saw for Linnaeus's trip to Lapland in 1732, Walker also included references to the culture and customs of the local populations. Thus, not far from the Kelani River, Walker described an agglomeration of rocks, caves, and waterfalls with reference to a 'tradition among the natives that from the top of the largest and highest of these boulders of rock, a queen of Candy, in former days, precipitated herself, or was thrown by her husband'.[28] Referring to the sacred value of Adam's Peak to the local populations, she concluded that at every visit to the altar on the top, the faithful believed they had taken 'a step towards heaven' judging by the extreme happiness shown by the locals who accompanied them (probably also in anticipation of the hefty compensation they would receive, Walker added maliciously).[29]

While Walker was engaged in Ceylon (after the ascent of Adam's Peak in 1833, she and her husband explored the southwest of the island in 1837), a much more important journey in the history of science took place: the second voyage of the HMS *Beagle* from late 1831 to late 1836 to carry out hydrographic and geological surveys in different areas of the planet but, of course, famous above all for the participation of Charles Darwin. He had the opportunity to see the landscape and flora and fauna of South America, the Pacific and Australian islands, the islands in the Indian Ocean, and South Africa, thus developing and testing his scientific theories on the evolution of species.[30]

During the five years of travel, Darwin came into contact with the populations of the places visited along the route, analyzing their customs and culture. However, the scientist's reports particularly express his fascination for the pristine natural settings he found when making landfall. As in the cases of Spallanzani and Walker, Darwin also had a certain fascination for wild mountains. He provided a rapturous account of the ascent to a cone that was a day's navigation north of Cape Tres Montes (Taitao Peninsula, Chile), which reminded him of the Sugar Loaf in Rio de Janeiro. The Chilean setting was a 'lofty weather-beaten coast, which is remarkable for the bold outline of its hills, and the thick covering of the forest'. Although not very high (1,600 feet), the top of the cone proved difficult to reach, a few days before Christmas 1834 he wrote:

> It was a laborious undertaking, for the sides were so steep, that in some parts it was necessary to use the trees as ladders…. In these wild countries it gives much delight, to gain the summit of any mountain. There is an indefinite expectation of meeting something very strange, which, however often it may be balked, never failed with me to recur on each successive attempt. Every one must know the feeling of triumph and pride which a grand view from a height communicates to the mind. In these little frequented countries there is also joined to it some vanity, that you perhaps are the first man who ever stood on this pinnacle or admired this view.[31]

We note an element of extreme tourism in the passage, that is, the quest for a specific emotion during the journey, in this case the satisfaction of overcoming the challenge posed by the difficult ascent. In this respect, we can see an analogy with Spallanzani's desire and determination to get as close as possible to the lava of Vesuvius and to Etna's crater. In the case of Darwin, there was also the enthusiasm of visiting a place supposedly never touched by Western civilization.[32]

The impact of the cliffs and peaks on Darwin also emerges in other passages of his account. For example, in November 1835, he and the local guides went up the Tia-auru valley in Tahiti. Skirting the watercourse that flowed through it, Darwin initially passed through the woods that flanked it, glimpsing the high Tahitian peaks in the distance. But the landscape soon changed:

> The valley soon began to narrow, and the sides to grow lofty and more precipitous. After having walked between three and four hours, we found

the width of the ravine scarcely exceeded that of the bed of the stream. On each hand the walls were nearly vertical.... These precipices must have been some thousand feet high: and the whole formed a mountain gorge, far more magnificent than any thing which I had ever before beheld.[33]

In this case, Darwin's enthusiasm for heights was not expressed so much as a desire to scale an inaccessible hill as by experiencing them from below, venturing into a deep gorge between imposing walls in the shadow of the peaks that dominate Tahiti. In addition, it was almost as wild as the Cape Tres Montes in Chile. Continuing up the valley, Darwin and his guides found branches of the stream that were inaccessible in some stretches due to water-falls. All but one:

The sides of the valley were here nearly precipitous; but, as frequently happens with stratified rocks, small ledges projected, which were thickly covered by wild bananas, liliaceous plants, and other luxuriant produc-tions of the tropics. The Tahitians, by climbing among these ledges, searching for fruit, had discovered a track by which the whole precipice could be scaled.[34]

By Darwin's own admission, the route was 'very dangerous' in some places, especially since it was necessary to cross stretches of bare rock, where the use of ropes was essential. However, the trip also offered some breathtaking views, as that from a ledge that passed behind an impressive waterfall drop-ping from hundreds of feet above their heads.[35]

The adrenaline-pumping experience in the Tahitian valley therefore had elements necessarily connected to the local population, their knowledge of the area, and tools such as ropes to move in greater safety. Nature remained almost untouched, but in this case, Darwin considered the interaction and participation of the local population to be fundamental. At the same time, Darwin's fascination for wild nature and its contrast with anthropogenic set-tings clearly emerges in another passage, where he comments on his stay in the city of Salvador, in today's State of Bahia, Brazil, and explorations of the surroundings in August 1836:

I was glad to find my enjoyment of tropical scenery had not decreased even in the slightest degree, from the want of novelty.... It must be remem-bered that within the tropics, the wild luxuriance of nature is not lost even in the vicinity of large cities; for the natural vegetation of the hedges and hill-sides, overpowers in picturesque effect the artificial labour of man.[36]

This contrast articulated by Darwin allows us to introduce a second cate-gory of scientific traveler. These are travelers interested in domesticated nature, agroecosystems, and local specialties, containing the seeds of today's rural and gastronomic tourism. They include naturalists who contemplated some elements of the agricultural sector, rural society, and the countryside in their surveys and real agriculturists whose main interest was studying rural practices.

The European context was characterized by the coming of age of agricultural science, which in the second half of the eighteenth century and the first decades of the nineteenth century achieved increasing autonomy from the other scientific branches and technical knowledge on which it had originally depended. The British agricultural revolution, the circulation of Linnean thought, the French *agronomes* and ascendent physiocracy, and the technocracy of Napoleonic Europe were both stimuli for and outcomes of the interest of experts, intellectuals, and politicians in the agriculture sector. This spread to countries hitherto characterized by a certain backwardness in economics and agriculture, such as Spain and some parts of Italy.[37]

Naturally, the new interest of European experts in agriculture also expanded to other continents. Alongside exquisitely scientific journeys, and sometimes intertwining with them, travels were undertaken by those with a keen interest in the economic potentials of local crops, animal husbandry, and derivative products. Experts frequently focused on rural environments. Their investigations might take place in the local countryside, seeking information on lesser-known agricultural practices. More often, as we have seen for naturalistic trips pure and simple, the travelers focused either on the European hinterlands or on rural areas on less explored continents. Let us examine some examples.

The first example is Alberto Fortis, naturalist, translator, and journalist *ante litteram* operating in the second half of the eighteenth century and in the early years of the nineteenth century, therefore during both the Old Regime and the Napoleonic era. Born in Padua, he lived in Venice, Naples, Paris, and Bologna, and since 1795 he was a fellow of the Royal Society of London. He made numerous scientific trips to Italy and the Balkans often together with fellow naturalists from various backgrounds, publishing his own reports, which were translated into the main European languages and read internationally. Let us consider the trip he took in the summer of 1773 in Dalmatia, then part of the Republic of Venice. The journey had a practical purpose, carried out at the bequest of the Senate of Venice to study the fishing conditions.[38]

Although Fortis was mainly oriented toward the natural sciences and in particular geology in his travels, he was always very interested in the interaction between humans and the Dalmatian environment, thus evaluating the state of agriculture, animal husbandry, and related activities, including fishing. His published reports contain numerous descriptions and reflections on the rural landscape, although they are often framed in a rather critical judgment of the backwardness of that region, with rare exceptions.[39]

An example of the rural potential in Dalmatian lands and the scarce exploitation that was made of it is provided by the following passage:

> At the foot of the mountain of Crisiza lies the beautiful valley of Dizmo, which affords good pasture, and the soil is not unfertile; it has about ten miles of circuit, altogether surrounded by hills. It is not cultivated, though it might easily be so; but the Morlacchi are very far from understanding good agriculture, and indeed they know little of any kind of it.[40]

The term *Morlachs* designated a community living in the Dinaric Alps. Despite often expressing negative judgments on the progress of their agricultural and manufacturing activities, Fortis was very interested in studying them as part of a general 'discovery' of the Balkans by European Enlightenment scholars. The experience gained over the years of contact with the Morlachs, the emotional bond with their culture and their traditions, and the esteem toward some Dalmatian intellectuals are elements that emerge in Fortis's writings.[41] It should be added that the entire Dalmatian agricultural sector was indeed backward, so much so that even in the Napoleonic era, the Kingdom of Italy tried to invigorate it through the work of Vincenzo Dandolo, general commissioner from 1806 to 1809.[42] While appreciating the beauty and fertility of the natural landscape in which the Morlachs lived and admiring much of their culture, Fortis criticized what he saw as an inability to exploit natural resources in an appropriate manner.

A similar line of judgment also emerged in many other passages, for example, one in which the backwardness of local agriculture was combined with the lack of strategy in fishing, despite the territory providing ample means in both areas. He considered the case of Lake Prokljan not far from Šibenik and Skradin (in the past known as Scardona):

> The lake of Scardona is altogether surrounded by hills of gradual ascent, and susceptible of the best cultivation; but they lie for the most part abandoned. Nor is the fishing practiced in a better manner than agriculture in these parts, although they are frequented by tunny, and other smaller emigrating fish. The fishermen attend to little else besides the daily provision of suitable fish, for the tables of the inhabitants of Sibenico and Scardona.[43]

He followed this with a rather detailed list of fish, revealing in a naturalist's lexicon the perspective of a gourmet, with references to the 'delicious morsels' and 'equal goodness' of some species, noting, of course, that the catch was poorly organized. The same taste for fish characterized the entire Primorje, a coastal and insular area with a Mediterranean climate between the islands of Pag and Sibenik and on the coast of the Seven Castles. In that case, Fortis paused to describe the abundant local catch of mullet, which was cleaned and salted. This time the naturalist was more sparing in his criticisms. Particularly in the city of Makarska they used to extract the roe pouches and dry them in the sun, producing yet another 'delicious morsel' (bottarga). Fortis also compared the mullet preparations made in Dalmatia with those produced in Comacchio and other parts of Italy, or in Greece, comparing their size and taste.[44]

In the description of Split, located along the coast about ninety kilometers southeast of Skradin, Fortis once again indulged his attraction to tasty fish, referring to an anecdote about the Roman emperor Diocletian, who was native to the area and died in Split. According to some sources, the emperor 'renounced the pleasure of commanding almost all the then known earth to eat quietly his bellyful of these fishes' in his beautiful villa. But Fortis admitted:

> I do not know if Diocletian was as great a lover of fish, as he was of herbs; but I believe that Spalatro [*sic*], without any motive of gluttony, must then have been a delicious habitation; and, to strengthen this belief, I imagine the neighbouring mountain to have been covered with ancient woods, which, in our times, by its horrid bareness, reverberates an almost unsupportable heat in the summer days.[45]

Fortis foreshadows a current issue in the tourism industry, especially in nature-based tourism, rural tourism, and most notably in ecotourism: to what extent the interaction of human beings with the environment can be considered sustainable.[46] Fortis certainly did not refer directly to the matter; he was carrying out a scientific and economic investigation to map the resources of Dalmatia. However, his comments on the beauty of the landscape and the quality of the food clearly reveal that—among his criticisms—he found reasons for personal satisfaction during his trip. Finally, considering the emphasis on fishing and fish-based nutrition in his 1773 expedition, we might even venture it was an *ante litteram* form of pescatourism, experiential tourism immersed in the fishermen's routine. Is Fortis's approach to travel really so different from the account a tourist might give after a week of rural tourism?

Spanish clergyman Antonio José Cavanilles, professor of botany and director of the Madrid Botanical Garden from 1801 to 1804, provides another agroecosystem travel case study. Trained in Paris from 1777 to 1789, he was both a renowned taxonomist and later an agricultural scholar and correspondent for numerous European scientific institutions. During his scientific career, Cavanilles always combined his interest in taxonomic botany with that in botany applied to the enhancement of agriculture, food production, manufacturing, and economic productivity in general. This attitude

emerges in many of his written works and in the interest with which he analyzed the environment encountered on some of his naturalistic journeys.[47]

In particular, between 1791 and 1793 he made some trips to the Valencian
region on a royal commission with the task of studying the local botany,
geography, agricultural practices, and traditions. As in the case of Fortis,
Cavanilles was not undertaking scientific journeys to particularly distant
and wild destinations. Moreover, the city of Valencia was his birthplace,
although he had moved away many years earlier. His descriptions reveal
experiences that closely resemble forms of rural tourism *ante litteram*, combining his personal reflections and the opinions of experts consulted during
his travels, providing a very articulate picture of his travel experiences.[48]

For example, he was fascinated by the forested mountains of the Tinença
de Benifassà, now a national park near the border separating the Comuni-
dad Valenciana from Catalonia and Aragon. He described the coexistence
of wild and domesticated nature, which was a particularly interesting feature
of the landscape. He was struck by the steep, barren slopes of the clefts in
the mountains, which he described as 'tall, impregnable castles'. The
Montnegrell was prominent among its neighboring hills and 'black with
many pines'. At the same time, Cavanilles appreciated the pleasantness of
the cultivated fields in the Bellestar area and around the local monastery,
which adorned a harsh yet seductive landscape, 'where you can observe
nature almost in its pure state'.[49]

Farther south, near the town of Castellón de la Plana, Cavanilles visited
the village of Vilafamés and the surrounding countryside. Describing the
village, he focused on its position perched on the side of a hill, likening the
steep streets to tormenting stairways and criticizing the fact that the only
source of water was inconveniently far from the center. However, he appreciated the magnificent church. What a tourist enjoying twenty-first-century
comforts might experience as a picturesque and very beautiful Spanish
village, the perhaps overly severe Cavanilles saw as an impractical arrangement. However, his enthusiasm was awakened when he visited the nearby
countryside, where olive groves, vineyards, carob groves, lupine fields, canta-
loupes, vegetables, and some of the tastiest watermelons in the whole of
Valencia were thriving and productive. Beekeeping and honey production
were also widespread; Cavanilles commented approvingly, although he preferred what he had seen in the valleys of Biar and Albaida, much farther
south toward Murcia, which were rich in rosemary, lavender, thyme, and
other aromatic herbs whose flowers allowed the bees to produce a more fra-
grant honey.[50]

Still in today's province of Castellón de la Plana, but much further south
than Vilafamés, Cavanilles visited the lands along the Rio Palancia, admiring woods, fields, vineyards, olive groves, orchards, and population centers
such as Segorbe and its neighbors. He was favorably impressed by the

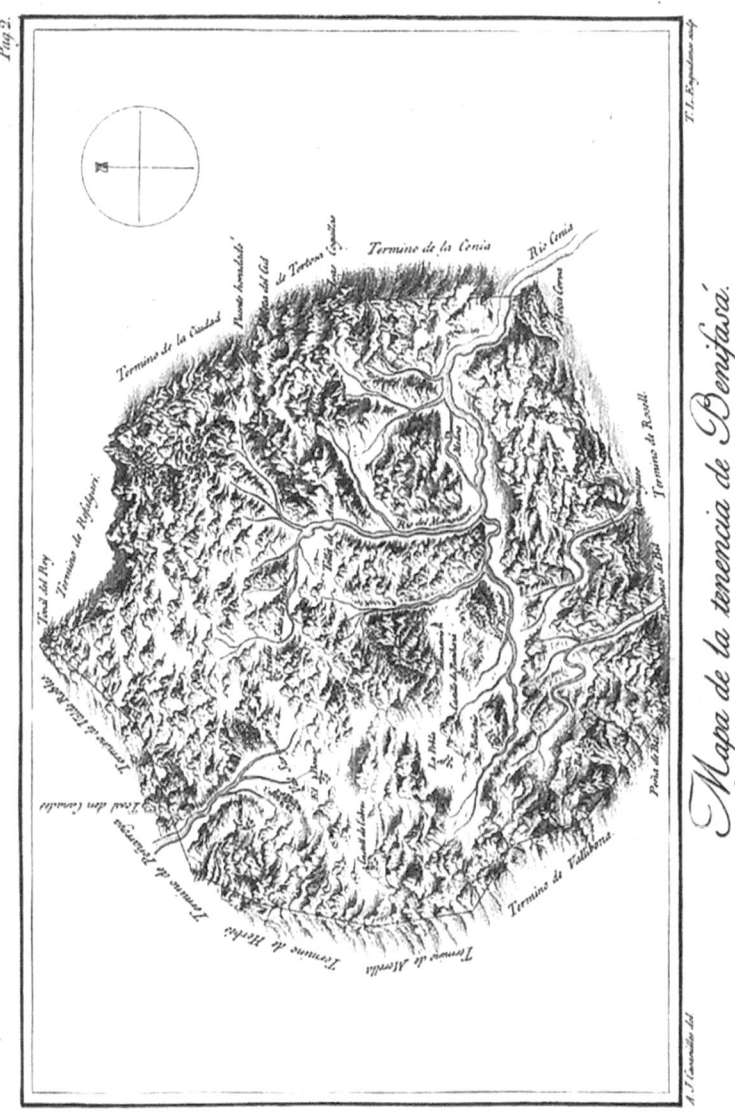

Figure 1.1 Map of the Tinença de Benifassà. Antonio José Cavanilles, *Observaciones sobre la historia natural, geographia, agricultura, población y frutos del Reyno de Valencia: volume I* (Madrid, 1795), p. 2.

Pag. 5.

Vista del Monasterio de Benifazá.

A. El Monasterio. B. Bellestar. C. Bel.

Figure 1.2 View of the Tinença de Benifassà with the local monastery and Bellestar. Antonio José Cavanilles, *Observaciones sobre la historia natural, geographía, agricultura, población y frutos del Reyno de Valencia: volume I* (Madrid, 1795), p. 5.

Charterhouse of Vall de Crist, where the monks cultivated their lands with great care—indeed he felt they could represent an agricultural model. He admired the numerous sources of crystalline water in the Viver area which, generally originating in high places, flowed downward 'in splendid water-falls, endowing the area with canals and coolness'. Observing those lands from a height, he admired the spectacle offered by the numerous streams of water and the varied shades of green of the crops, an incomparable bucolic view.[51]

The famous naturalist Alexander von Humboldt also offers us descriptions of the interaction between humans and nature, with numerous references to agricultural activity in contrast with unspoiled nature. In his case, these are much wilder lands than those explored by Fortis and Cavanilles in the Mediterranean. Humboldt traveled to the Americas from 1799 to 1804 and through Asian Russia in 1829. The first journey was a personal and almost entirely self-financed initiative. The Spanish Crown granted approval to Humboldt and the French botanist Aimé Bonpland (1773–1858) to pre-pare for their departure by studying materials produced by the Madrid scientific community and to embark on the *Pizarro*, a corvette leaving from Galicia. Humboldt's second journey was the initiative of the Emperor of Russia, who financed a mining expedition to seek platinum (the naturalist no longer possessed the same economic fortune he had enjoyed at the turn of the nineteenth century).[52]

Humboldt's botanical, geological, and mineralogical knowledge and related scientific pursuits intertwine with an emotional description of the travel experience. Already in his *Essai sur la géographie des plantes* (1805), his attitude as a scientist emerges clearly in his comment on equatorial vegetation, a belt humans were unable to tame. It hid the sun from their eyes and left only the ocean and rivers free. But he comments on its 'wild and majestic character', two elements that distinguished this type of environment, where any effort of human culture was overwhelmed by the luxuriant vegetation.[53] For him, nature—especially the plant kingdom with its enormous variety—had a powerful effect on the imagination, senses, and creativity of human beings.[54]

We may discern an affinity between Humboldt's vision and that of Darwin during his stay in Salvador in the summer of 1836, when he described the contrast between unspoiled nature and the anthropic landscape in Brazil. Moreover, Darwin added a series of provocative questions for those who were not learned travelers in the tropics:

Who else from seeing a plant in an herbarium can imagine its appearance when growing in its native soils? Who from seeing choice plants in a hothouse can magnify some into the dimensions of forest trees, and crowd others into an entangled jungle? Who when examining in the cabinet of the entomologist the gay exotic butterflies, and singular cicadas, will associate with these objects, the ceaseless harsh music of the latter, and the lazy flight of the former—the sure accompaniments of the still, glowing, noonday of the tropics.[55]

For both Humboldt and Darwin, the pristine nature of the tropics was something extraordinary that shaped the human beings who lived in it every day—rather than vice versa—and was inconceivably beautiful for those who had not seen it in person. However, in his travel reports Humboldt also made numerous references to the pleasantness of rural settings, where human intervention was a more decisive factor. The following examples are excerpted from the accounts of the expedition across the Atlantic and into the Americas with Bonpland.

On the way, stopping off on the island of Tenerife, in the Canaries, he described the abundant cultivation of dates, banana trees, sugar cane, prickly pears, olive trees, vines, European fruit trees, and cereals. He pointed to the successful acclimatization of many exotic crops. He described the cultivation and use of taro (*Colocasia esculenta* (L.) Schott), from whose rhizomes the populace obtained a nutritious starch: taro is still widely used today in Canarian cuisine, especially in vegetable stews. Alongside the crops, he showed his admiration for the luxuriant tropical flora, 'the abundance of plant forms revealing the majesty of regions near the equator'.[56]

He also penned a glowing description to the rural landscape of La Victoria and neighboring lands in Venezuela, then in the Viceroyalty of New Granada. While the inhabited center was characterized by bustling streets full of shops, the surrounding lands were peaceful and idyllic. To the west, the splendid valleys of the Aragua River were covered with gardens, fields planted mainly in wheat, farms, small towns, and clusters of wild trees. To the south and east, a huge massif hid the savannahs of the Los Llanos hinterland from view. Here too, alongside the anthropogenic elements of the landscape, the natural elements stood out, as was the case in the Canaries.[57]

Regarding agricultural history and tradition, Humboldt stressed, among other things, the value of agriculture in the calendar and pantheon of pre-Columbian peoples, often venturing beyond the confines of the classical field of natural sciences. He referred, for example, to the figure of Centeōtl, whom he called 'the Mexican Ceres', the Aztec deity who protected the crops and to whom the ancient natives offered human sacrifices. He sought analogies with the corresponding divinities in the 'western nations', analyzing their symbology and role in calendars.[58]

We therefore note in Humboldt's interest in the rural landscape and the bounty of cultivated lands an attraction to wild nature and the 'darker aspects' of the lands he visited, an element not found in Fortis's reports on the Balkan coast (which was more urbanized than many of the American areas). Humboldt would often linger in his reports on the most exciting aspects of the trip. One example is his description of Tequendama Falls, in Colombia, 'tremendous scenery' to which local populations—overwhelmed by its grandeur, shifting rainbows, and column of mist that could be seen at a distance of five leagues—attributed divine origins. He also related the crossing of the Chambo River, in Ecuador, near Penipe in June 1802, on a rope bridge 120 feet long and 7 or 8 feet wide, crossed one at a time, quickly, without stopping, bent over forward and swallowing one's fear.[59]

We close this part dedicated to the experiences of scientists with Enrico Hillyer Giglioli, an Italian English zoologist and anthropologist who took a journey that touched Latin America, the Asia-Pacific, and Australia in the years 1865–1868, when he was in his early twenties. He traveled on the frigate *Regina* from Italy to Uruguay and then on the steamship *Magenta* with his mentor Filippo De Filippi (1814–1867), director of the Zoological

Figure 1.3 Rope bridge near Penipe over the Chambo River, Ecuador. Alexander von Humboldt, *Researches Concerning the Institutions and Monuments of the Ancient Inhabitants of America with Descriptions and Views of Some of the Most Striking Scenes in the Cordilleras! Written in French by Alexander de Humboldt and Translated into English by Helen Maria Williams: volume II* (London, 1814), plate 13.

Museum of Turin and supporter of Darwinism, who died of infectious hepatitis in Hong Kong during the expedition. The objective of the trip was mainly of a commercial nature: the commander of the *Magenta* had the task as plenipotentiary of the king of Italy to make agreements with China and Japan for the liberalization of sericulture at a critical time for Italy, in its infancy as a unified state. Giglioli and De Filippi instead participated as members of a classic scientific mission to collect animal and plant samples, as well as information on the environments and populations encountered along the way.[60]

Unlike Fortis and Cavanilles and more similar to Humboldt and Darwin, Giglioli was not primarily focused on the rural context but associated the experience of unspoiled nature with the analysis of the culture and economy of the populations encountered. Here, too, we may appreciate the intertwining of natural landscapes and anthropogenic environments in the travel report published in Milan in 1875 with the patronage of the Ministry of Agriculture, Industry and Commerce and dedicated to the deceased De Filippi.

Regarding tamed nature, he gave an admiring description of the Japanese rural landscape, characterized by the 'scrupulous precision of the fields' and its 'truly florid and advanced appearance'. In addition to indicating the species cultivated in each area in the different seasons, describing the agricultural techniques of the Japanese, and outlining the activities in the rural centers, he paused to describe some idyllic landscapes. He visited Japan in July and August 1866, combining direct observations with information on seasons and places he could not reach personally collected from colleagues and locals, allowing him to reconstruct a general description of the land and landscape. The hills north and west of Tokyo Bay were colored yellow in early April by the 'blossoms of the field mustard that fill the air with a pleasant fragrance'. When rice was harvested in November, there was a solemn celebration among the Japanese peasants, with songs and dances accompanied by crude musical instruments and celebrations of Inari—a deity of agriculture and rice, among other things—to ensure the fertility of the fields for the new harvest.[61]

Walking in the summer countryside near Ajiro, on the Izu Peninsula, he took a road among small rice paddies and climbed a hill dotted with hydrangeas. In the shade of the Masson's pines, his eyes were greeted by numerous tombs on which flowers, bowls of water, rice, salt, and scented sticks had been carefully placed. He then went down to a village where the houses were screened by gardens of bamboo, palm trees, oranges, roses, wisteria, and many other plants. In addition, some of his traveling companions visited the Atami baths, with hydrotherapy establishments offering all the comforts. Indeed, they were frequented by international personalities such as the former British plenipotentiary minister Rutherford Alcock in 1861.[62]

The descriptions of Giglioli's travels contain features of not only rural tourism but also cemetery tourism and spa tourism.[63] The Japan trip put him in a strongly anthropic context that contrasted distinctly with other descriptions, such as that of the Brazilian forests visited in January 1866. It is true that he also described the urban bustle of Rio de Janeiro, but deep in pristine nature, the sensation was completely different:

> The sky was blue and the sun rays penetrating through the green fronds made the drops from last night's rain sparkle like diamonds on the leaves. The copious evaporation that immediately begins after a rain in these regions is truly extraordinary.... When you move into the shade of those towering trees, you have a sensation that is very similar, but much more powerful, to that you have entering a greenhouse in the winter: the atmosphere is hot, heavy, humid, and filled with a heterogeneous mix of smells from flowers and decomposing vegetable matter.... It is a strange, voluptuous, profound sensation, but it cannot be described to someone who has not experienced it.... No sound reached my ear in the forest; a deep silence reigned everywhere; the large butterflies (*Morpho*) fluttering lazily here and there seemed to be beings from another world; only a few birds interrupted the solitude of plants.[64]

In Brazil, it was the charm of unspoiled nature and, as Darwin also repeatedly described, sensations beyond what people in common walks of life could experience that struck Giglioli. It was a dreamlike and disturbing dimension at the same time, while the Japanese countryside was dreamlike and soothing with its regularity and pastel colors. The contrast that the attraction to both kinds of landscape, natural or domesticated, had on his soul is tangible in this observation:

> In Japan we may say that very little land capable of producing some useful plant remains uncultivated, and the botanists who have visited this country have often criticized the way that crops occupy almost all land, leaving on a tiny portion for their research into indigenous plant species that are less useful to non-botanist humanity.[65]

However, he admitted that what his colleagues and predecessors expressed was 'a bit exaggerated' and that wild flora was present there too. For him, the experience in Japan had been extremely pleasant and he would later add: 'I leave Japan with a heavy heart; it will often be in my thoughts, and my most cordial wishes for its fortune and happiness'.[66]

1.3 Humanists

The subjects we analyze in this section are characterized by nature travel experiences not motivated by scientific purposes. It therefore includes writers, philosophers, memorialists, and artists, who left above all for personal growth and sometimes due to other contingencies, managing to collect their notes and publish them successfully.

Some of them possessed some knowledge of natural sciences and agricultural science. As we will see, this was the case of the philosopher and orientalist Volney and the botanical illustrator Marianne North. However, research in these scientific fields was not the cause that led them to travel, even though elements of them may be found in the descriptions they gave of wild and domesticated nature, although in a much less pronounced form than in the accounts of naturalists and agriculturists examined earlier, as is only logical.

We also introduce in this section an analysis of the cases with a table that summarizes the profiles of the travelers we consider, showing their nationality and travel areas analyzed herein.

A particular example of a traveler belonging to the humanities category is the Frenchman Constantin-François Chassebœuf de La Giraudais, Count de Volney. He was a philosopher and an active supporter of the French Revolution but not of its excesses; he was later an opponent of Napoleon's absolutism, from which he nevertheless benefited (such as receiving the title of count). He was also an important connoisseur of the history, culture, and society of the Ottoman Empire, as well as of the Arabic language.

Table 1.2 Travelers analyzed in Section 1.3. Our elaboration

Name	Life	Nationality	Travels analyzed in Chapter 1
Count de Volney	(1758–1820)	French	North Africa and Middle East
La Rochefoucauld-Liancourt	(1747–1827)	French	North America
Mary Wollstonecraft Shelley	(1797–1851)	British	Europe
Marianne North	(1830–1890)	British	North America, South Africa, and Indian Ocean
Giovanni De Riseis	(1872–1950)	Italian	East Asia and Pacific Ocean

He developed his knowledge in this area with in-depth studies but, above all, by visiting Egypt, Palestine, and Lebanon between 1783 and 1785. In the period 1795–1798, he visited the fledgling United States. It should be noted that as a young man Volney studied medicine; despite the wealth of scientific knowledge he commanded, his travels were not dictated by the study of fauna, flora, or minerals.[67]

The account of the journey of 1783–1785, published in 1787 with the title of *Voyage en Syrie et en Égypte*, contains a number of passages expressing his interest in the landscape. A key aspect of this case study, however, is that Volney is not always enthusiastic about direct contact with the natural environment. He provides both positive and negative criticism, discussing the pros and cons. To better clarify the 'interlocutory' approach in his report, let us take some critical examples.[68]

The first is the description of the Egyptian landscape as an eyewitness. He rightly praised Egypt for its great fertility along the Nile River, the great variety of its products, and its advantageous position for trade. This aspect reflected his interest in agricultural improvements. In the following years, he would be appointed by the French government as general director of agriculture and commerce for Corsica and would acquire lands on the island to conduct agricultural experiments.[69]

His positive observations on Egypt were accompanied by negative ones: the climate, the great number of scorpions and insects, especially flies 'such that one could not eat without risk of swallowing one'. Furthermore, he stated:

> Besides, no land could be more monotonous, always a barren plain as far as the eye could see, always a flat, uniform horizon; date trees on their thin stalks, mud huts along the roads. Never a richness of landscape or the variety of objects or the diversity of sites that fills the eyes and the

mind with revitalizing scenes and sensations.... There are neither clear streams nor cool tussocks nor secluded spots; there are no valleys, hills, or towering rocks.[70]

This trenchant judgment clearly illustrates Volney's dislike for arid or relatively infertile landscapes, an opinion similar to that expressed a century later for some parts of southern Africa by the painter Marianne North, which we analyze later. However, as we will see in this and the next chapter, this sort of landscape—in whatever region of the earth it was located—was not always a completely negative experience for all travelers. Those undertaking to explore them more thoroughly often found much to appreciate.

For example, the Italian geologist Giovanni Battista Brocchi was in the service of the viceroy of Egypt in the 1820s and was commissioned to conduct a thorough exploration of the hinterland for mining purposes. We will return to the details of his travels in Africa and the Levant in Section 1.4. Of course, he, too, affirmed in his travel comments that the landscape in the middle of the desert was rather harsh, and like Volney, he preferred the fertile Nile Valley. However, his attitude was different. Volney recognized the importance of the Nile for agriculture and said he understood the 'religious respect' historically accorded it by the Egyptians but added with the usual arrogance that a European, thinking of much clearer waters, would have smirked upon hearing talk of the beauty of the Nile, seeing it instead as turbid, muddy, and yellowish. Brocchi, by contrast, dwelled on the fascinating aspects of this 'majestic river', on the crops and villages along the banks that he admired on his upriver boat journey, on the peace that reigned in the starlight at night, regretting that there was no full moon to bathe the palm groves in its light.[71]

Volney's critical spirit reappears when he describes the Levant, both its arid and green areas. For example, he described with some admiration the surroundings of Tripoli, in present-day Lebanon, as 'orchards where the prickly pear grows wild, the mulberry is cultivated for silk, and pomegranate, orange, and lemon trees for their fruits, which are most beautiful to behold'. However, in his opinion, the large amount of water that was channeled during the summer to irrigate these fruit groves and the poor air circulation due to the spurs of Mount Lebanon that almost entirely surrounded Tripoli resulted in considerable stagnation that caused 'epidemic fevers' in the summer months. So paradoxically, the system supporting domesticated nature combined with the local morphology leading also to negative effects: despite the traveler's eye being pleased by the variety of crops, the greenery around Tripoli also had a negative aspect. This was alleviated, Volney continued, only at Tripoli Marina (El Mina), a seaside town with a harbor located not far from the city but with a naturally airier climate.[72]

In the United States, Volney was impressed by the vast forests, but even here his reaction to nature was not enthusiastic. He describes the forests in the first pages of the *Tableau du climat et du sol des États-Unis d'Amérique* published in 1803, which contains Volney's experiences on his American journey in 1795–1798.

> For a European traveler, and especially for a traveler such as myself who is accustomed to the barren regions of Egypt, Asia, and the shores of the Mediterranean, the salient feature of American soil is the wild aspect of the forest which is encountered immediately at the ocean shore and grows thicker and thicker inland.[73]

The comparison was therefore both with urbanized and massively cultivated Europe and with the deserts of Egypt and the Levant. It certainly emphasized Volney's lack of sympathy for hot, dry landscapes. However, he did not seem to experience American nature, green and often almost wild, as a refuge from a largely manmade world such as Europe.

In 1796, he journeyed from the mouth of the Delaware to the Wabash River and from there to Fort Detroit before returning eastward to Albany, New York. The following year, he traveled from Boston to Richmond, Virginia. Volney wrote that on such journeys, 'I was hardly ever able to walk even three miles on cleared land'. The roads—or 'dirt tracks', as he characterized them with some disappointment—were shaded by woods or tall forests ruled by silence and monotony, with soil that was now arid, now marshy, many trees uprooted by old age or storms lying on the ground to rot. He also detested the swarms of biting flies, gnats, mosquitoes, and other insects (for which Volney had a particular aversion, as already demonstrated in Egypt). He certainly did not find 'the charming effects that novelists writing in our European cities dream of'.[74]

Going from Fort Detroit to Albany in 1796, Volney saw Lake Erie and Niagara Falls. His description was very different from Kalm's of nearly fifty years earlier and from that of another French traveler, the Duke of La Rochefoucauld-Liancourt (whom he mentioned and to whom we shall return shortly), who saw them shortly before Volney. It cannot be said that Volney was not impressed, but he immediately dismissed any romantic enthusiasm by writing that the other French travel writers had dedicated themselves above all to the description of the 'magnificent spectacle' of the falls, while he would have concentrated on the 'topographical features', of which the falls were merely the outcome. In the following pages, he drew on the natural studies from his medical education and took a more rational and aseptic approach to the landscape and his travel experience, curiously, perhaps more rational and aseptic, despite his humanist profile, than in the memoirs of the scientific travelers discussed earlier.[75]

Volney did include a few passages where he expressed a positive inclination toward American nature. For example, when exploring the land near Niagara Falls and along the St. Lawrence River, he arrived at the edge of an escarpment where the trees thinned out, 'suddenly an immense horizon' opened up before him with the view of Lake Ontario, which looked like a sea. However, such comments are infrequent in his *Tableau*.[76]

His general analysis of the territory was relatively detached, appraising the economic benefits it could provide but often peppered with rather critical comments. Volney described with some accuracy the plant species that made up the great American forest that was 'almost everywhere', flanked by large bodies of water, mountain ranges, and prairies. His list of plant types was organized by region on the basis of soil type and quality and whether it was productive or not for agricultural purposes. And he quite frequently complained about the capricious sky and weather that in one day brought 'Norwegian ice, African sun, and all four seasons of the year'.[77]

Volney therefore offers us the example of a traveler who gained a certain experience with natural landscapes over the course of his life, passing from Egypt to the Levant to the United States. From his accounts, it appears that he knew how to analyze these environments with a certain precision, combining his personal experiences with the studies of naturalists and the accounts of other travelers. It appears that the first of our humanistic case studies— albeit here 'tainted' with a background in the sciences—rife with criticism and Europocentric references, does not represent a true passion for nature travel. As suggested earlier, Brocchi gave us a more nuanced view of Egypt than Volney; let us now consider Rochefoucauld's views on the United States.

François Alexandre Frédéric de La Rochefoucauld-Liancourt was a French nobleman, politician, the founder of the École nationale supérieure d'arts et métiers, and an advocate of vaccines. A supporter of the monarchy, he was forced to emigrate from France during the Revolution. After a short stay in England, he left Europe in 1794 and reached the United States in 1795, remaining there until 1797. Similarly to Volney's travels, Rochefoucauld's experience abroad was not based on scientific research or diplomatic or economic objectives. Beyond the political and institutional situation that had forced him to leave his country, Rochefoucauld's interest in North America was that of a man curious about the culture, society, and geography of those distant lands, not a naturalist engaged in a scientific expedition or an explorer or diplomat on a mission.[78]

Let us analyze the description he gave of Niagara Falls. In his travel narratives, Rochefoucauld admits that Niagara Falls were 'one of the principal objects of our journey, and which I had long desired to see'. He and his traveling companions, led by an American lieutenant on a boat made available by a Major Pratt of Fort Erie together with six strong soldiers at the oars, went up the Welland River in June 1795 from Fort Erie to Fort Chippawa (both in present-day Ontario, Canada) where horses awaited them.

> Each stroke of the oars brought us nearer to it, and our attention being entirely turned to discover the foam, and hear the noise, we took but little notice of the banks of the river, which, on the side of Canada, are tolerably settled, of the uncommon width of its channel, or the majestic course of its stream.[79]

Approaching the falls, the emotion felt by Rochefoucauld and his traveling companions is well expressed, the waiting with bated breath, the anxious desire to hear or see the first signs, the attention focused wholly on the goal. This differs significantly from the much more detached description that Volney would give some time later in his *Tableau*. Rochefoucauld's description of the falls themselves is precise, objective, and technical, but unlike Volney's, it is also enlivened by expressions such as 'grand spectacle' and 'this stupendous cataract'. He describes the water dropping into the void and changing color depending on the way the sunlight hit it, passing from 'dark green' to 'foaming white'. In this, he evokes the description provided by Humboldt of the Tequendama Falls, in Colombia, cited in Section 1.2.[80]

The sight of Niagara Falls deeply impressed Rochefoucauld, to the point where he used it as a litmus test for his other experiences in the United States. For example, in 1796, he found himself admiring the confluence of the Shenandoah and Potomac Rivers in West Virginia: a 'beautiful' and 'majestic' display of the fast-moving turbulent waters of the former meeting those of the latter at a right angle in the shadow of the Blue Ridge Mountains. Rochefoucauld admitted: 'The scene is grand; it deserves to be viewed, and is worthy of the admiration of travellers who delight in the magnificent operations of nature'. But he had not felt the enthusiasm that he had expected based on the accounts of other travelers who had seen the two rivers meeting, such enthusiasm could only be elicited (and he clearly reiterated this) by admiring Niagara Falls.[81]

The accounts of Volney and Rochefoucauld provide two points of view on the journey in natural settings. They are partly opposed but both assess the experiences with a critical eye, making comparisons to journeys in other places at other times or to those contained in the memoirs of other travelers. We also observe that, unlike Rochefoucauld, Volney's experience of nature tourism *ante litteram* was not the most pleasant, as evidenced by the high number of negative criticisms relating to the landscapes he encountered, in part driven by an inevitable European-centric evaluation of an extra-European context. However, he was anything but narrow-minded, as demonstrated by his interest in the history and culture of Egypt and the Levant. But he shows little inclination to abandon himself to the experience of nature and the unknown in his *Voyage* and *Tableau*, which convey a certain predilection for agricultural landscapes, although we cannot go so far as to interpret this as an embryonic form of rural tourism.

Mary Wollstonecraft Shelley, author of the novel *Frankenstein*, provides a different story of a journey 'in nature'. Mary and her husband, the poet Percy Bysshe Shelley (1792–1822), left England for the continent in March 1818 for a variety of reasons. Not only did they want to escape Percy's creditors, but the couple was also accompanying Mary's stepsister, Claire Clairmont, to Italy. Claire was taking her daughter to stay with her biological father, Lord Byron, who lived in Venice. The Shelleys were also motivated by creative impulse: the trip through Europe to Italy would take them into novel milieux and hopefully bring new poetic and literary inspiration, as had been the case in the summer of 1814 when they took a brief trip through France, Switzerland, Rhineland, and Holland and, in the summer of 1816, on Lake Geneva. The journey of 1818–1822 was tragic in many ways. Mary lost two young children and her husband in Italy. It would be almost twenty years before she returned to Italy with her son Percy Florence and some of his friends on two trips that also included other European countries in 1840 and again in 1842–1843, when the group went down to the Amalfi Coast.[82]

Mary Wollstonecraft Shelley, a woman of letters and an attentive observer of society, was attracted to the great urban centers of Europe, especially the ancient cities of Italy that were obligatory stops on the Grand Tour, such as Venice, Florence, Rome, and Naples. We read this in her accounts of the Italian sojourn of 1818–1822 and her European travels in 1840 and 1842–1843 (those of the 1840s were published as *Rambles in Germany and Italy* in 1844). She also expressed a fondness for natural settings, whether wild or bucolic, which suited the writer's melancholic and turbulent character. This is especially clear in the earlier trip to Lake Geneva and the Alps she took in the summer of 1816.

Together with her future husband (they would marry in December) and her half-sister Claire Clairemont, the writer left the Alpine town of Chamonix to see the source of the Arveyron, the stream that emerges from the vast glacier known as Mer de Glace, on the Mont Blanc massif. She first traveled on the back of a mule and then from a certain point on foot. The Mer de Glace has been shrinking since the mid-nineteenth century, but on this trip, the writer noted that 'this Glacier is increasing every day by a foot, closing up the valley'. She was quite impressed by the glacial landscape generally and by the implied danger, as she wrote in her journals:

We came to the [source] which lies like a stage surrounded on the three sides by mountains and glaciers – we sat on a rock which formed the fourth-gazing on the scene before us – an immense Glacier was on our left which continually rolled stones to its foot – it is very dangerous to go directly under this ... we see several [avalanches] – some very small others of great magnitude which roared and [smoked] – overwhelming every thing as it passed along and precipitating great pieces of ice into the valley below.[83]

She also had the opportunity to move upland from Lake Geneva to visit the surrounding Alps. The writer particularly appreciated mountaineering and naturalistic adventures, savoring a certain sense of intrinsic danger, while entrusting herself to guides to keep her safe.

For the trip to the source of the Arveyron and the Mer de Glace she related a story told by their guide, according to whom two Dutch travelers had ventured into a glacier cave without a guide and had stupidly fired a pistol, 'which drew down a large piece on them'. A few days earlier, she had also visited the Cascade de Chedde near Passy, in Haute-Savoie, on the back of a mule. Here she described the impressive drop of over 200 feet and 'the rain of the spray' that soaked her clothes. She added suggestively: 'This cataract fell into the Arve which dashed against its banks like a wild animal who is furious in constraint'.[84]

Wollstonecraft Shelley's interest in nature was therefore a significant component of her travels. She also expressed it many years later, in early summer 1840, in a very different and less wild context, when she left France via Thionville to visit the Prussian countryside:

The country after this grew more varied and pleasant, but the villages deteriorated dismally. They were indescribably squalid. The dung before the doors—the filth of the people—the wretched appearance of the cottages, form a painful contrast, which too often presents itself to the traveller, between the repulsive dwellings of man and the inviting aspect of free beautiful nature, all elegant in its forms, delicious in its odours, and peaceful in its influence over the mind.[85]

Albeit in a more domesticated form, a conception of nature and its beneficial effect on human beings emerges from these lines that is very similar to the one we analyzed for Humboldt and Darwin in the tropics. Already on her first trip to Italy with her husband, she enjoyed country life and the peaceful coexistence of human beings and nature. Leaving Milan in early May 1818, she traveled by carriage through the Emilian countryside, describing it as 'pleasant', 'fertile', and 'picturesque'. In the Parma area, she dwelled on the horizon sketched out by the Apennines and on the cornfields dotted with trees 'up which the vines are trained and then festooned from one tree to another'. After crossing to the southern side of the range she commented that 'the south side of the Apennines is more picturesque and clothed with chestnut woods'.[86]

Similarly, in the summer of 1820, she received a visit from a friend at the Bagni di San Giuliano, a spa in the Pisan countryside where the writer and her husband were staying at that time. While they chatted about literature and news from England, rural life unfolded outside their window amid cheerful noises, like the grunting of pigs brought to be sold at the fair.[87]

She provided a more eloquent and detailed representation of the natural aspects of the countryside that could be seen from Lake Zurich in July 1840, on which she had her son Percy Florence take her in a boat. However, as in the Pisan countryside, here, too, the dominant feature was the comfort and reassurance of the agricultural ecosystem:

> This lake is not extensive nor majestic as that of Geneva, with its background of the highest Alps; nor as picturesque and sublime as Lucerne, with its dark lofty precipices and verdant isles; but it is a beautiful lake, with a view of high mountains not very distant, and its immediate banks are well cultivated, and graced by many country houses. After dinner, I went out in a boat with P—, by ourselves; ... we crossed the lake, which is not wide at this point, and returned again by moonlight.[88]

An image of sweet and accommodating nature also emerged in other passages of the *Rambles*. For example, in July 1842 Wollstonecraft Shelley admired the 'romantic' hills around the Bavarian town of Bad Brükenau, covered with dense beech forests, a type of old-growth community now on the list of UNESCO World Heritage Sites in several areas of Europe (including Germany), representing the ancient flora of the continent. Of course, the forests around Bad Brükenau featured trails and benches 'constructed for the convenience of the visitors', so they were a far cry from the wildernesses explored by Humboldt in Latin America or Walker in Ceylon. Her quiet strolls cannot be compared to Darwin's daring adventures in Tahitian gorges or Spallanzani's perilous explorations on the slopes of Italian volcanoes.[89]

Similar tones are found in Wollstonecraft Shelley's description of the entry into Italy from the Splügen Pass in mid-July 1840, comparing it to the arrival in paradise after old age and death:

> All Italian travellers know what it is, after toiling up the bleak, bare, northern, Swiss side of an Alp, to descend towards ever-vernal Italy. The rhododendron, in thick bushes, in full bloom, first adorned the mountain sides; then, pine forests; then, chestnut groves; the mountain was cleft into woody ravines; the waterfalls scattered their spray and their gracious melody; flowery and green, and clothed in radiance, and gifted with plenty, Italy opened upon us.[90]

Wollstonecraft Shelley also experienced the pleasure of hospitable, soothing nature when she visited Capri in June 1843, a completely different landscape from Bavaria. In Capri, she devoted herself to sea tourism. She visited the Blue Grotto by boat, enjoyed meals and evenings in the company of others on the terrace of her accommodation, let herself be captivated by the beauty of the night sky, and visited the ruins of Tiberius's Palace on donkey-back amid vineyards and Mediterranean scrub. The love she expressed for the Mediterranean region in her *Rambles* might elicit in our minds the roughly coeval paintings by the German artist Jakob Alt: soft colors, dreamy views, conciliatory nature.[91]

The author of *Frankenstein*—a novel conceived during her stay on Lake Geneva in 1816, immersed in rain and ghost stories—was also moved by the darker, more menacing aspects of nature, referring to 'jagged pinnacles and bare crags' during her Alpine adventures in the summer of 1816. And in her later travels, she acknowledged the more gothic aspects of her surroundings.[92]

Figure 1.4 Jakob Alt, *Die Blaue Grotte auf der Insel Capri*, watercolor, 1835–1836. Albertina Gallery, Vienna.

For example, the Höllental in the Black Forest struck her imagination when she visited it in July 1840. She wrote in the *Rambles* that some points could 'charm a painter' and others were 'almost beyond the reach of imitative art'. But she also admired the ability of the Germans to give 'the glory of spirit-stirring names to their valleys and their forests' (Höllental means valley of Hell). She described the valley in these terms:

> The Höllental is indeed a narrow ravine shut in by hills, not very high, but rocky and abrupt, and clothed in the rich foliage of majestic trees. In parts the ravine closes in so as to leave only room for the road between the precipice and the mountain-river ..., which now steals murmuring between its turf-clad banks, and now roars and dashes in a rocky bed. Jagged pinnacles and bare crags overhang the road; around it are strewn gigantic masses of fallen rock, but all are clothed with luxuriant vegetation, and adorned by noble woods.[93]

Her description of the approach to the city of Linz, Austria, at sunset in September 1842 is expressed in somewhat more rapturous tones. The city was nestled among wooded hills with the Salzburg and Styrian Alps towering on the horizon. The Danube 'swept under high precipices' and then continued with its 'glittering waves' that gave 'life and sublimity' to the landscape. The sky above was '[g]olden and crimson, the clouds waited on the sun, now dazzling in brightness; and now, as that sunk behind the far horizon, stretching away in fainter and fainter hues, reflected by the broad river below'. In a similar description, the anthropic element of the city was absorbed by sublime and breathtaking nature, whose beauty, accentuated by the sunset, attracted the attention of travelers. The view of Linz was complementary, adding 'much to the harmony and perfection of the landscape'.[94]

Our next case study regards Marianne North. She was a renowned English painter of landscapes and especially of botanical subjects. She spent her life traveling around the world. In particular, after the death of her father and early traveling companion in 1869, she decided to visually document the flora of distant lands. Numerous naturalists, including Darwin himself, suggested particular countries for her work. In 1882, her gallery was opened at Kew Gardens, the main English botanical institution; the structure was designed by the Scottish architect James Fergusson, especially for her paintings.[95]

After the painter's death in 1890, her sister Janet Catherine, wife of the poet, literary critic, and erstwhile gay rights activist John Addington Symonds, edited Marianne's writings, which were then published in the volumes *Recollections of a Happy Life*. Although the autobiography, written almost entirely by Marianne, was further edited and revised by the publisher—altering the author's narrative style—with professional corrections of botanical terms, it appears to conserve a certain level of authenticity about the feelings experienced by the painter in her travels around the globe. Let us analyze some passages from the *Recollections* in which the natural setting of North's work on botanical paintings becomes a key component of what we would now term experiential tourism.[96]

In 1871, she visited the United States and Canada. She had the opportunity to cross beautiful Lake Winnipesaukee, in New Hampshire, in the autumn, aboard a steamer 'under the light of a great full moon'. During the crossing, she was especially struck by the countless islands that dotted the lake and the surrounding hills and mountains approaching 6,000 feet in elevation. She was able to admire the lake during the day, enjoying the fall foliage and concluding that the name of the lake, meaning 'Smile of the Great Spirit' in the language of the Abenaki (a still disputed translation), was most appropriate:

> The views by daylight were very curious, owing to the gorgeous colouring of the maples and sycamores; nothing but our most brilliant geranium-beds could rival the dazzling variety of reds and crimsons, and the blue Michaelmas-daisies made tiny pyramids of colour in the foreground, the white ones looking like miniature fir-trees loaded with snow.[97]

North's description conjures a pristine natural landscape in our minds. However, an anthropogenic presence had already been established and would continue to develop in the following century, making Lake Winnipe-saukee a tourist attraction combining nature and recreation, and well served by communication routes.[98]

In addition to the steamer on the lake, North had traveled comfortably to Alton from Boston by train with an acquaintance. Fresh water, apples, pears, Isabella grapes, popcorn, and cakes, were available in the car for passengers. Moreover, her hotel accommodations on the lake were not lacking in comforts:

> From six till nine there was an endless breakfast, twelve to three dinner, and five to eight supper, with an enormous list of dishes for each individual to choose from and order for himself—piles of hot cakes like pan-cakes, pumpkin-pies made with treacle and eaten with Cheshire cheese, a huge fish called holibat [*sic*], and chowder and ice-cream.[99]

She also referred to the fact that 'the season was nearly over', thus suggesting that there was already a defined periodization in the flows of visitors to the lake.[100] We are therefore a long way from the unspoiled and extreme nature encountered by Darwin in the southern hemisphere some 35–40 years earlier. Both for the more advanced era and for the geographical area in which it was located, North's travel experience featured elements of nature tourism and rural tourism more similar to the contemporary concept of such types of tourism, with convenient means of transport, hospitality infrastructure, and an abundant supply of food. In the following pages, we shall examine other examples from source material on North to ascertain how common these characteristics were in her travels.

Between August 1882 and June 1883, North traveled in Southern Africa. Her destinations were British colonies, where she was able to take advantage of modern means of transportation and the hospitality of English officials and naturalists. Here she did show a greater engagement with wilderness areas than she had on Lake Winnipesaukee, although always through the lens of modern urban amenities, and not always with pleasure.[101]

For example, traveling by train from Aberdeen to Port Elizabeth, now Gqeberha in the Republic of South Africa, she admired 'a new world of vegetation' in the hills that passed by outside the window. She was struck by the many aloes, prickly pears, spekbooms, 'great cactus-like euphorbias', and other species, which she described in their variety of colors and shapes. She also glimpsed clusters of huts in the clearings, which she described in a somewhat strange way with folkloristic terms. She recognized the 'superb dignity' of certain natives, who 'never even turned their heads as the train passed'. Here, too, she was able to enjoy Western comforts: the city of Port Elizabeth was 'full of life and work, very clean and neat, with an excellent hotel'. The previous stretch of the route, from Beaufort West ('a lively little town and a good hotel') to Aberdeen, had been more daring and more exposed, which explained her relief as she approached Port Elizabeth. It had taken her two days to travel this 120-mile portion on horseback with a guide with heavy drinking problems. The road was good, 'but it was all desert, covered with small scrub, strange dwarf euphorbias, and miniature plants'. 'Bleached skeletons, tall stalking-birds, and a few deer', along with a varied array of mirages, were the only elements to break the monotony.[102]

Later, after almost three weeks of travel from Marseille, on October 13, 1883, North landed on the island of Mahé, in the Seychelles, remaining in the archipelago until the beginning of 1884, when the lack of nutrition, intense work, and quarantine on Long Island due to a smallpox epidemic caused her to have a nervous breakdown. At that point she returned to England.[103]

The beginning of her stay, at least, was pleasant. In the town of Victoria on Mahé, where the English commissioner Barkly resided, buildings and streets intermingled with lush vegetation of coconut palms, vanilla festoons, *Barringtonia*, and others. On the afternoon of her arrival, after being greeted 'most kindly' by Barkly and his wife, she crossed the low pass to the other side of the island, which she described as 'exceedingly beautiful', with cinnamon and cloves that 'were both growing luxuriantly' and vanilla everywhere 'trained on espaliers'. Her outings on Mahé and expeditions to the other islands in the following weeks continued in the same vein. North traveled between islands on boats provided by English notables and on foot or litter on the islands, guided by local functionaries to the various spots for her work.[104]

The interaction between North and the non-British locals was also important. She gained an important perspective from her acquaintance with a certain Madame Chocolat, 'a very black lady' with whom she found various affinities, such as the desire for solitude and tranquility. During a walk together, Madame Chocolat pointed out the painter to some friends, happily exclaiming 'This is a real Englishwoman!' revealing how exoticism is a two-way street. In any case, in general, while admiring the flora and beaches dotted with 'shells, corals, and blue crabs, a world of wonders', North could not help noticing that 'all the inhabitants seemed on the verge of starvation, and living on credit' and recorded and sometimes criticized local activities, such as their method of fishing.[105]

We can therefore conclude that North's trips varied greatly depending on the adequacy of transportation and amenities. While there was always a significant anthropic presence in the destinations she visited, in some her contact with nature was more direct and intensive than in others. Moreover, the *Recollections* reveal, if not true sensitivity—she is critical in many passages and exhibits bias toward all things not English—at least a sincere interest in the indigenous peoples, their traditional customs, their economic resources, and perhaps their social problems.

We close this section with the case study regarding the Italian nobleman Giovanni De Riseis. In addition to being a major landowner due to numerous inherited properties, he held important political positions in his later years, including podestà of Naples and senator of the Kingdom of Italy starting in the 1930s, after being provincial councilor of Chieti from 1902 to 1912. However, we do not include him among the working travelers in Section 1.4 for a number of reasons. First of all, the trips he made to different parts of Asia took place at the end of the nineteenth century when he was still young and had not yet assumed political office; they were thus not associated with diplomatic missions or political purposes. They were also not business trips related to his role as landowner and entrepreneur. De Riseis traveled for the pleasure of getting to know new countries, new cultures, and new environments. The factor aligning him with the humanist camp is the fact that he produced a great deal of writings about his experiences.[106]

Il Giappone moderno was published in Milan in 1895 and immediately reprinted the following year. De Riseis had spent the last months of 1893 in Japan as part of a much more extensive trip involving a long stay in the United States and then in Burma (now Myanmar) and India. On his own admission, the description he gave of the experience was not intended to be 'an analytical study' but 'simple notes taken *in loco* and momentary impressions'. We shall analyze the passages describing the interaction between human society and nature rather his genuine attraction to the primordial element.[107]

The book begins with a description of the San Francisco Bay contrasting natural and anthropic elements. He juxtaposes the 'warm October sunset', 'the calm waters of the bay gilded with the sun's rays', and 'the beautiful blue sky' with large numbers of Chinese people embarking on a ship operated by the Occidental and Oriental Company, 'long lines of Celestials disappearing into the dark hatches of ocean liners, as if an immense tomb were opening before them'. Many of these were elderly people seeking to return to their homeland to die there, explains De Riseis, adding:

> There were frightening faces in those groups embarking with us: yellow, withered, reduced nearly to skeletons. The terrible suffering and unspeakable horror all Chinese feel for the sea was clearly written on every one of those doubly tormented faces.[108]

The description seems unwittingly to presage the somber tones of Joseph Conrad's novel *Typhoon*, written a few years later and published in 1902, in which a steamer loaded with returning Chinese workers encounters a tropical typhoon in the South China Sea. However, De Riseis's crossing from America to Japan was also characterized by pleasant images, such as a young whale that 'darted along like a giant dolphin' in a 'race with the boat' or 'the languid tropical evenings' during which the boat 'seemed to be plying phosphorescent waves amid thousands of flying fish'. However, after Hawaii, the weather worsened and the ship fell prey to long storms, with 'furious winds' that forced the passengers below deck and 'high pitched whistling' of the rigging at night. However, the ship continued on its course undaunted, making 300 miles a day.[109]

Contrast is the element that distinguished Japan in the eyes of De Riseis. He considered Yokohama to be an 'excessively noisy and hectic commercial city', the only refuge being the Bluff District on a hill south of the city, historically the abode of diplomats and foreigners, and where the Italian Consulate was located. Here the avenues of cherry trees (albeit 'bare and unfruited' when De Riseis saw them during the cold season), the shady streets, and the thick woods that shielded the houses formed 'a truly ideal spot for the misanthrope who flees the company of his fellows to renew his spirit with the splendid view from ahigh in the unbroken silence of the fragrance woods'. Past the city in the foreground and the endless rice fields stretching northwards toward Tokyo, the view also encompassed the sea and Mount Fuji in the distance.[110]

The anthropic presence was a constant in urbanized Japan even on excursions De Riseis made far from the main cities with guides and travel companions, seeking contact with nature. It is true that when descending from the mountain town of Hakone to Atami on the coast he was able to admire dark, dense woods, Mount Fuji reflected majestically in the waters of Lake Ashi, and take in a superb view of the succession of mountains leading up to the Jikkoku Toge Pass (Ten Province Pass), whipped by the impetuous wind. As walked through the woods on red-earth paths toward the city of Ōzu, he savored 'a great peace, perfect tranquility'. Equally pleasant was the quiet 'green basin' of Kamakura, described as 'a jumble of hills that descend from on high to dip into the waters of the sea', or the slopes 'speckled with the most beautiful flowers' on the island of Enoshima. But traces of humanity were everywhere: the peasants here and there engaged in daily tasks, rice fields, huts and villages, temples, votive statues, boats, the railway, the shops, and the villas of Japanese aristocrats and foreign diplomats on the coast.[111]

This was certainly not the pristine wilderness of Darwin's Latin America or Tahiti or Walker's Ceylon in the 1830s. It was also different from the Japan seen by Giglioli twenty-seven years earlier, even though the zoologist had focused on the incipient anthropization of some areas of Japan.[112] It is also true that De Riseis belonged to a different category of traveler from that of scientists, who were professionally oriented toward wild nature and the study of ancient local traditions. In this respect, De Riseis was much closer to Wollstonecraft Shelley: a leisure traveler, who would publish pleasant yet noncommittal reports, at ease in society and urban centers yet also willing to spurn them and go off in search of allegedly pristine nature, provided it is easily accessible and practicable.

1.4 Working Travelers

After considering the journeys of scientists and humanists, let us now analyze the description of nature by those engaging in a 'business trip'. The travelers are laypeople or clerics who found themselves experiencing natural and agricultural ecosystems while traveling for professional purposes. The motivation for the trip is not scientific research or interest in different cultures, but more material or strategic purposes, which may be economic, political, or even pastoral. While this might seem to be an incongruous grouping at first sight, it is justified precisely by the common element of a practical, concrete, planned goal, even if it regards the care of souls or missionary work.

Of course, it could be objected that many naturalists and agriculturists taken into consideration in Section 1.2 on scientific travelers could have equally been included in this section if they were engaged in expeditions that examined the territory not only for research purposes but also to collect material information on the potential economic applications of flora, fauna, ecosystems, and local practices. However, it is necessary to distinguish the role played by these travelers during the journeys considered. Many of these scientists accompanied expeditions led by professional explorers and were not in charge of the expedition; for example, the expedition in which Darwin participated on the HMS *Beagle* was led by vessel Captain Robert FitzRoy, who played a decisive role both in adapting the *Beagle* for the voyage and in choosing who should participate.[113]

Another type of involvement is that exemplified by the case of Brocchi, contracted by the viceroy of Egypt, which we have already had the opportunity to mention and which we analyze in more detail shortly. It represents a criterion for including some scientists and technicians among working travelers: their greater degree of responsibility within the expedition as the ones directly in charge, resulting in a different, more technical and pragmatic, approach to the travel experience.

Cases like Wollstonecraft Shelley and De Riseis in Section 1.3 on humanist travelers, however, stand out sharply from the profiles in Section 1.4. The two did not travel 'for work' but out of curiosity and interest (and, in the case of Wollstonecraft Shelley, actually to get away from economic issues). Also from Section 1.3, Marianne North produced paintings of botanical species during her travels that were then exhibited on several occasions. Nevertheless, the later use of these works was not the main purpose of North's travel experiences. She had begun to travel and paint as a way to cope with her grief for the death of her father, with whom she had traveled in continental Europe, the Levant, and Egypt, and because sketching and painting in watercolor were two of the few activities available to wealthy spinsters in Victorian societies.[114]

Here, too, we propose a table that outlines the profiles considered, showing their nationality and the travel destinations analyzed in the section. As we will learn in the following analysis, some of them—and particularly Brocchi—had an exquisitely scientific-naturalistic background that might qualify them for placement among the naturalists and agriculturists of Section 1.2. However, we reiterate that the goal of the trips and, more generally, their professional activity are the criteria for placement in this third category of ecosystem travelers.

The first case is that of Alexis-Marie de Rochon (Abbé Rochon), an astronomer in the French Navy who traveled extensively in the Indian Ocean. In the late 1760s, he participated in the mission of Captain Jacques de Grenier to find a faster sea route from Mauritius (then French) to India and had the opportunity to visit the coasts of Madagascar in 1768. He was a fundamental member of the crew as an official astronomer who had the task of calculating and recording the coordinates of the voyage. He served in the same role on a subsequent voyage, in the first half of the 1770s, with similar purposes in the same area, this time under Captain Yves Joseph de Kerguelen.[115]

Table 1.3 Travelers analyzed in Section 1.4. Our elaboration

Name	Lived	Nationality	Travels analyzed in Chapter 1
Alexis-Marie de Rochon	(1741–1817)	French	East Africa and Indian Ocean
Alessandro Malaspina	(1754–1810)	Italian	Pacific Ocean
Giovanni Battista Brocchi	(1772–1826)	Italian	Northeast Africa and Middle East
Giovanni Battista Balangero	(1849–[1902?])	Italian	South Asia and Oceania

Rochon's impressions of Madagascar's nature are illustrated in several passages of his *Voyage à Madagascar et aux Indes Orientales*, published in Paris in 1791, with the English translation published the following year in London. He accompanied that account of his experiences with a reconstruction of the island's history based on secondary documents and sources.

The astronomer's 1768 visit to Madagascar was in the company of Pierre Poivre (1719–1786), intendant of the islands of Mauritius and Réunion (at the time Isle de France and Isle Bourbon), whose task was to find new and interesting plant species to enrich the botanical garden at Monplaisir. Recently established, the garden would soon acquire a lofty reputation for the great variety of species—most providing either spices or pharmaceuticals—earning the nickname of *Jardin du Roy à l'Isle de France*. As Rochon himself reported in his book, Poivre had the merit of having spread nutmeg, cloves, breadfruit, and a variety of dry-cultivated rice from Cochinchina to the French colonies. The administrator therefore most closely aligned with the category of economic botany, which involved scientific explorations and exchanges of species on the global level.[116]

Rochon wrote of the forests of Madagascar:

The forests contain a prodigious variety of most beautiful trees, such as palms of every kind, ebony, wood for dyeing, bamboos of an enormous size, and orange and lemon trees.... These numerous trees and shrubs are surrounded by a multitude of parasitic plants and vines. In these forests may be found agaric and mushrooms, the colors of which are lively and agreeable, and which have exquisite savor.... All the forests of Madagascar abound with plants unknown to botanists, some of which are aromatic and medicinal, and others fit for dyeing.[117]

These lines convey the fascination of luxuriant, mysterious nature, and the engagement of the senses. However, also clearly stated is the interest in the practical uses of the species and the potential of applying science and technology to this natural reservoir. Features of modern science tourism are therefore also recognized. In another passage, the description had a more romantic air, referring both to uncontaminated nature and to the domesticated nature of the agroecosystems:

> The traveller, who, in the pursuit of knowledge, traverses for the first time wild and mountainous countries, intersected by ridges and valleys, where nature, abandoned to its own fertility, presents the most singular and varied productions, cannot help being often struck with terror and surprise on viewing those awful precipices, the summits of which are covered with trees, as ancient, perhaps, as the world. His astonishment when he hears the noise of immense cascades, which are so inaccessible for him to approach them. But these scenes, truly picturesque, are always succeeded by rural views, delightful hills, and plains where vegetation is never interrupted by the severity and vicissitude of the seasons. The eye with pleasure beholds those extensive savannas which afford nourishment to numerous herds of cattle, and flocks of sheep. Fields of rice and potatoes present, also, a new and highly interesting spectacle.[118]

It is interesting to note how this passage contains some of the aspects that most fascinated also the scientific travelers whom we discussed earlier in this chapter: wild nature characterized by mystery, surprise, and awe; waterfalls, whose grandeur is the embodiment of that awe; domesticated nature, where a pleasant, hospitable landscape reveals the success of science and technology in promoting the economic exploitation of natural resources.

Still within the framework of economic botany, Rochon's *Voyage* includes a description of shrubs and plants that he found in northern Madagascar then brought to the intendant Poivre. For each species, Rochon reported the local names and provided very vivid descriptions. We shall consider two of these. For the *malao-manghit*, a sort of nutmeg, visuals set the tone: Rochon described it as a broad tree with a straight trunk, brown bark, and white wood; its sap was 'white and milky' but became 'red as blood' when exposed to air. For clove nutmeg (*Cryptocarya agathophylla* van der Werff), it was more the sense of smell to be involved: the essential oil extracted from its leaves was redolent of cloves, cinnamon, and nutmeg; Indian cooks used it in ragouts, preferring it to any other type of spice. The description of these visual details and sensory aspects thus complemented the other passages of the text on Madagascar focusing on the landscape.[119]

Under the guidance of General and interpreter La Bigorne, Rochon visited the area around Antongil Bay 'and with him I saw those astonishing quarries of rock crystal, the masses of which are so enormous as almost to surpass belief'. His interest in crystals and minerals derived from the fact that he was also a scholar of optics and an inventor of useful instruments in that field. Other indicative references emerge from the *Voyage*. He claimed that the mountains of Madagascar contained 'enormous blocks of rock crystal', that iron deposits of excellent quality were scattered throughout the island, and that mountains in the north and south had 'in their bowels' a great abundance of valuable fossils and minerals. At the coastal town of Foulpointe (Mahalevona), Rochon's attention was caught by stretches of sand, 'which appears to have experienced semi-vitrification' and which was mixed with 'stones of a soft friable nature' and 'an infinite number of small fragments of natural glass'. Minerals were therefore a part of Rochon's overall interest in Madagascar, uniting scientific-economic aspects, awe before the wonders of nature, and full sensory engagement.[120]

The second case study regards the voyages across the Atlantic and Pacific Oceans by the Italian naval officer Alessandro Malaspina from 1789 to 1794. Before the expedition, Malaspina participated in several naval missions in the service of the Crown of Spain and the Real Compañia de Filipinas, including a circumnavigation of the globe. Strengthened by his naval career and political contacts in Spain, he proposed an expedition with specialists from various scientific fields to deepen the knowledge of the colonial territories in Latin America and the Philippines. The journey he planned should have been another circumnavigation of the globe, but the corvettes *Atrevida* and *Descubierta* did not go beyond the coasts of Asia and Oceania. A great deal of information was collected about populations, cultures and economies, natural resources, the environment, and the topography of both the Spanish colonies and other lands. Malaspina carefully chose the scientists to bring aboard and meticulously organized most other aspects of the expedition.[121]

We analyze some passages of Malaspina's travel diary to understand how much attention this traveler with a military background and motivated by purely economic and political objectives dedicated to the variegated extra-European ecosystems that he would encounter on the voyage. In his diary, Malaspina did not dedicate the same space and the same descriptive accuracy to nature that we find in Rochon or in other travelers. In fact, attention is more focused on the technical aspects of maritime travel and on the cultural and economic characteristics of the companies encountered along the way. However, interesting observations emerge in several passages.

For example, in May 1790, the *Descubierta* corvette was off the Chilean coast and brought Malaspina to the archipelago of the Desventuradas Islands, which is difficult to reach and today devoid of civilian settlements. The Italian navigator painted a rather barren picture: 'all appear equally bleak and steep-to', without even seabirds or seals which, instead of taking advantage of the isolation of the archipelago, 'seemed to have been driven away by the bleak appearance of these islands'. Despite this, the members of the expedition showed an interest in those strange islands. They observed the steep coasts and their layers of different colors depending on the composition of the rock, as well as the sparse vegetation that could be seen from the ship.[122] In truth, the severe judgment of Malaspina toward the archipelago is in contrast with the importance it has for marine biodiversity. Indeed, it has been the subject of various scientific expeditions and has been recently included in the Nazca Desventuradas Marine Park created by the Chilean government.[123]

In July 1791, the expedition was in the Gulf of Alaska, and Malaspina recorded a rather sparsely populated coastal environment in terms of fauna (some seals and very few birds), but the lands east of Cape Suckling were 'covered with a pine forest as thick, unbroken and luxuriant as those we had seen around Port Mulgrave', with 'moderately high snow covered hills'.[124] In describing the coast near Cabo Nodales (now Point Martin), Malaspina noted that 'all its shore is sandy, but beautifully forested a short way inland'.[125] The two corvettes then dropped anchor off Mount St Elias—the fourth-highest peak in North America after Mount Denali, Mount Logan, and Pico de Orizaba—of which Malaspina was able to admire 'the noble natural architecture'.[126] These comments were in a clear minority compared to the technical and aseptic data related to navigation and geography. However, they also revealed the allure Malaspina felt for the harsh environment of Alaska, which still today has great interest for travelers from all over the world. In fact, coastal tours are organized in which both the peculiar environment and local cultures play a central role.[127]

When the expedition reached Sorsogon in the Philippines in mid-March 1792, Malaspina and his traveling companions were thrilled. He considered first of all the port where they had landed 'one of the most beautiful that nature has created' for the breadth of its bay and the right depth of the waters that allowed easy access to a large number of boats, but also for the wealth of good fish, as well as anthropic elements such the surrounding towns 'which can supply all necessities'. Furthermore, the area presented numerous other points of interest, especially for the naturalists of the expedition: 'the lush vegetation, the two volcanoes in sight, the proximity of these terrains to the more distant areas watered by the muddy Pasig [River] in the vicinity of Manila'.[128] Malaspina's notes therefore revealed a nature that became interesting because it was useful to human beings, both in a practical sense (a convenient bay for landing and all the provisions a traveler might need) and from the point of view of scientific interest (botanical and geological aspects to explore in the area).

Our next case study is set in the 1820s. Giovanni Battista Brocchi worked for the *Wali* of the Ottoman province of Egypt, Mehmet Ali (1769–1849). The context was one of great industrial, economic, and military renewal that would lead to the rebellion of the *Wali* himself against the sultan in the 1830s. Having gained solid experience as a geologist and mineralogist on several scientific journeys in Italy and the post of inspector of the Council for Mines under Napoleonic rule and during the first years of the Restoration, in the 1820s, Brocchi was hired through an Italian agent for the *Wali* named Giuseppe Forni, a chemist and a director of a gunpowder factory in Cairo who was temporarily in Milan. Brocchi's task was to conduct explorations in the Egyptian hinterland to locate ancient abandoned mines dating back to the Pharaonic era and possibly to direct the restoration work. Brocchi arrived in Egypt in early autumn 1822 with a contract that would have bound him to make two other exploratory voyages after his first trip to the Egyptian hinterland, which lasted until mid-1823: one in the Levant from mid-1823 to mid-1824 and a second trip in the Egyptian hinterland as far as present-day Sudan from March 1825 to September 1826, when he died in Khartoum, probably of dysentery.[129]

Brocchi always showed great interest in the flora, fauna, and sociocultural aspect of the lands he visited, both on his Italian travels and in explorations in the service of the *Wali*. The mission stipulated in his contract remained the core of his experiences in Egypt, the Levant, and Sudan; however, along the way, he constantly recorded details about the surrounding environment—both wild nature and local populations—in what was published posthumously in several volumes starting in 1841 as a *Giornale delle osservazioni fatte ne' viaggi in Egitto, nella Siria e nella Nubia*.

We have already considered his perception of the Nile as a 'majestic river' when he sailed up it in a felucca, the peace of the nocturnal landscape, and his description of villages and crops on the banks. Brocchi also recorded several scenes of the rural life of peasants and cattle farmers. He describes what he considered a 'highly singular' scene:

> A herd of buffalo on the left bank of the Nile crossed over to the right bank by swimming, carrying the young herdsmen on their backs. They made the long trip almost completely immersed in the water, with only their muzzles protruding. The boys either sat on the backs of the buffalos or stood, holding their clothes bundled on their heads. With equal ease an ox also passed by, guided by a cowherd who was also swimming, with a group of gourds under his chest to better keep him afloat.[130]

While folkloristic scenes like these on the Nile produced little more than an astonished record by the foreign traveler, Brocchi found greater involvement with the peasants on Mount Lebanon, praising their industriousness despite obstacles of various kinds and evaluating their relationship with the environment with the eye of an economist. He was impressed by the technical skill of the Lebanese peasants in building terraces on the mountainsides and perfecting mulberry cultivation and silk processing. Indeed, their abilities were in high demand at Wadi Tumilat, close to the Nile delta, where they were summoned as skilled labor to manage extensive mulberry plantations.[131]

While many of the Lebanese peasants had learned to read and write, they lived in conditions of extreme poverty and few of them owned land. The landowners, on the other hand, tended to be illiterate—not unlike the Italian nobles, commented Brocchi—and often had to rely on their subordinates even to read the names of the months. Meanwhile, the local representatives of the Turkish authorities were not champions of public utility projects. While a spring had been channeled in the countryside near Beit ed-Dine, bringing irrigation and power for mills to the surrounding area, it was Brocchi's opinion that it was only built because the local emir thought it would make his palace and gardens more pleasant.[132]

Brocchi also looked with an idyllic eye on what he perceived as a return to nature regarding the custom of many Lebanese peasants and people of humble status to go barefoot and bare-legged even in winter. We may assume that it was a question of poverty, but Brocchi interpreted it as a tradition inherited from their forebears contrasting with European habits. The framing of the rurality in a classical context and the juxtaposition of technical descriptions with georgic reminiscences was a very common practice in the treatises, journals, and accounts of experts of the early nineteenth century.[133]

In other cases, the analysis of the human–ecosystem relationship focused more on local food, making many comparisons between European tastes and those of the lands visited. For example, when Brocchi arrived in the Sennar region of Sudan, he noted the absence of poultry farms, while he recorded the cultivation of small and tasty onions in some areas. The diet was based on milk and sorghum or millet bread fashioned into thin wafers, with extensive use of a sour drink still very common in Sudan called *merissa*, obtained from fermented sorghum. Tamarind was used both in a drink that 'takes the place of our lemonade' and as a condiment for dishes 'when you need something slightly acidic, where we would use lemon juice or vinegar'. He also noted:

> A very common food is a round squash about the size of a large pomegranate known as an *agiurum*, which is garnished in various ways. It is full of seeds as big as those of the carob that are sold roasted at the market, and sprinkled with a bit of salt you eat them with the peel, which many use as a substitute for bread, and so a great deal is consumed, and thus in a certain sense they are the equivalent to our potatoes.[134]

Also curious is Brocchi's description of 'urban nature' when he stayed in Jaffa, on what is now the Israeli coast, during the trip to the Levant in 1823–1824. Approaching the city, which stands on a hill, he and his traveling companions began to see a 'luxuriant vegetation' that was a relief after the dryness of the desert. To enter the city, they walked along an avenue bordered by flowering trees and prickly pears studded with fruit. Beyond, on both sides, there were pomegranate gardens laden with fruit, sycamores, figs, peaches, oranges, vineyards, and sugarcane fields; they admired the irrigation system that used water wheels. The beauty of these gardens reminded Brocchi of the garden of Alcinoo on the island of the Phaeacians, described in the *Odyssey*, and strongly contrasted with the aridity of Egypt, dominated by trees such as palms and sycamores, which were very scarce in Jaffa.[135]

The geologist had a nuanced perspective as compared to Volney's very negative and prejudicial view. Evidence of Brocchi's astonishment with the Egyptian and Sudanese settings is found in his account of the first time he saw mirages on a particularly arid stretch of the route. He wrote that the guides who accompanied him used to call them 'devil's water' in Arabic and explained that it was a sun-dependent phenomenon. Brocchi added that the phenomenon of the mirage was mentioned in the Quran: 'As for the disbelievers, their deeds are like a mirage in the desert, which the thirsty perceive as water, but when they approach it, they find it to be nothing'.[136]

When he reconnoitered the galena deposits near El Qoseir, on the Red Sea, in February 1823, Brocchi described a barren and harsh nature, but the reader can sense his fascination for it in spite of the technical vocabulary of the expert geologist and mineralogist at work. He moved in a setting of stony mountains that, depending on the area and the valley, ranged in hue from red to blackish to purplish-brown, with quartz veins of varying thickness. The route was traversed by ditches and ravines, which Brocchi curiously explored, escorted by the guides placed at his disposal by the local sheikh. The environment tended towards wilderness but bore the traces of ancient mining activities: wells, tunnels, abandoned galleries, ruins of houses, small shrines, and vestiges of writing on the rocks (hieroglyphs as well as Latin, Greek, Arabic, and Coptic characters).[137]

We conclude our discussion of Brocchi by noting the relative importance, with respect to the other accounts we have examined, of the human element in his experience of the natural scenery. His *Giornale* describes the elements of the environment with a keen spirit of observation and with the typical minutiae of an inspector with a solid scientific background, but he continuously referred to the anthropic element—past or present—in completing his reports. This feature was evident, albeit to a lesser extent, in the reports of another working traveler, Rochon in Madagascar, and logically also in those of the scientific travelers in Section 1.2 who were interested in agroecosystems. We therefore understand that the goal of the trip and the role of the traveler in it significantly affect the perception of nature and the different approaches to the wild dimension.[138]

Christian missionaries offer a further arena of analysis. There was a resurgence in missionary activity in the second half of the nineteenth century, partly in response to a new wave of colonial imperialism and driven by the erosion of church landholdings in the face of new, more material, and less spiritual models of life. The global missionary activity had been relatively subdued in the eighteenth century for reasons including the expulsion of the Society of Jesus—which had been the protagonist of the major missions in previous centuries linked to the great discoveries—from many states and their colonies starting in 1759 and its official suppression by the Holy See in 1773 (the Society of Jesus was restored in 1814).[139]

Giovanni Battista Balangero of Saluzzo, ordained a secular priest in 1871 at the age of twenty-two, was a missionary in Australia from 1872 to 1874 and in Ceylon from 1874 to 1885. He made numerous trips around the lands entrusted to his evangelization work during those years. Back in Italy, he was knighted and appointed spiritual director at the national boarding schools in Genoa, Venice, and Turin. Numerous young people with whom he was in contact were fascinated by his stories and urged him to print the material written during his missionary work and to narrate his travels. At first, he was opposed to this, believing it would be presumptuous and out of place to reveal the details of his work and his private life to the public. However, during the years of his mission, he had already shared his experience materially or in writing. For example, he had sent some specimens of Indian Ocean pearls to the Mineralogical Museum of the University of Turin. In addition, he had written an English epistolary report for the weekly *Ceylon Catholic Messenger* regarding his visit to Palestine during the return trip from Sri Lanka to Europe in 1885. He was then persuaded to write a book and in 1897 his collection of *Studi e ricordi di tredici anni di missione* went to press in Turin.[140]

Balangero immediately made a clear distinction between Australia and Ceylon in the introduction to his book. He considered the former to be a land of progress and wealth with flourishing agricultural, livestock farming, and mining sectors. Recent urbanization had wrestled land from the wilderness bringing economic development in its wake. Here, we note numerous similarities with the Japan of progress and urbanization described by Giglioli and De Riseis.[141] However, when the cardinal prefect of the Congregatio de Propaganda Fide (today Congregation for the Evangelization of Peoples) proposed that he go to Ceylon in 1874, Balangero proved his dedication to his missionary vocation. By his own admission, if he had wanted to 'make money and thereby live a life of ease', he would certainly have stayed in Australia.

He had the same ardor for Ceylon, describing it as a wild and bewitching land of natural wonders, clearly expressing the enthusiasm that Walker's more contained and scientific description had only hinted at in the early 1830s. These are the words used by Balangero at the beginning of the book, in which he paraphrased and interpreted the many names and attributes of the island, underlining how its nature was celebrated by different cultures and religions:

> Ceylon ... is a land whose history and language and monuments speak to us of ancient civilizations, a land that is like an immense tropical garden of unsurpassed beauty and richness ... ; it is so attractive and charming as to be commonly referred to as paradise on earth, a land the Brahmins call *Lanka*, meaning 'the glittering one', the Chinese the *Island of Jewels*, while for the Muslims it was the refuge of Adam and Eve after they were expelled from paradise; and finally a land seen by all Buddhists as the light and center of their world.[142]

Balangero expressed the same fascination for the nature of the island in his descriptions of botanical species. His diligent recording of scientific information, economic uses, and commercial value—elements typical of the nineteenth-century progress during the Second Industrial Revolution and increasing globalization—was embellished with flowery prose.[143]

He described the areca palm (*Areca catechu* L.) as 'most elegant' and quoted Hindu poets who compared it to 'a lightning bolt from the sky'. He called the talipot (*Corypha umbraculifera* L.) the 'most majestic and noble of the Ceylon palms'. He wrote that the breadfruit (*Artocarpus altilis* [Parkinson] Fosberg) and jackfruit (*Artocarpus heterophyllus* Lam.) trees 'are both beautiful'. Of course, he dedicated attention to how the local population used these and other species, sometimes making comparisons between native and European perspectives. He reported that breadfruits and jackfruits were an important element in the local diet, tasting of sweet potato when roasted but not very popular with Europeans. He wrote about the importance of the kitul palm (*Caryota urens* L.) as a source of sugar, wine, vinegar, and textile fibers 'thus feeding and employing an entire family of Ceylonese'.[144]

In his descriptions, Balangero made clear his sensitivity for Ceylonese culture and the knowledge of the land and resources of the indigenous people that he had gained over the years. We may guess that the poetic vein in his writing was influenced by his spiritual practice as well as by the literary style of the period. Yet there were also many episodes that shook his nerves during his long years on the island.

For example, Balangero often went to the church and attached missionary residence in Matara, a town on the southern coast of Ceylon. The buildings were located outside the town, and he used to stay there for a few weeks at a time. The church and residence were located on the banks of a river that he characterized as one of the 'most beautiful and poetic rivers that can possibly be imagined', whose waters flowed calmly between banks rich in 'stupendous vegetation'.[145] It was the Nilwala River, still home to a large population of estuarine crocodiles (*Crocodylus porosus*, Schneider, 1801), the world's largest reptile, in Sri Lanka. Balangero was well aware of both the beauty and the dangers of the environment in which he lived.[146]

Few crocodile attacks on humans have been recorded in past centuries, an increase only being recorded starting in 2000, resulting in retaliatory killings of crocodiles by humans and a real human–crocodile conflict. Scientists are studying the possible causes of the increase in crocodile attacks on humans. One hypothesis is the effects of sand mining, which significantly alters the river's physical, geological, and chemical characteristics, allowing crocodiles to travel much farther inland.[147]

Balangero warned his readers of the risk, mentioning an episode that had occurred during one of his stays in Matara. A family cook washing dishes by the river was killed by a crocodile who grabbed her by the head and dragged her into the water. Today, as in the days of Balangero, there are fences along the river to allow people to bathe and wash clothes safely. However, fatal attacks still occur due to carelessness in unprotected areas. In early September 2007, a man taking a dip in the Nilwala River was killed by a particularly large crocodile that was terrorizing the area. In early July 2020, a police officer was killed in Matara while trying to retrieve a cell phone that had fallen into the river.[148]

Let us now return to the dichotomy expressed by Balangero between Ceylon and Australia. He documented the wild and adventurous side of both, having experienced them firsthand during his travels. While describing Australia as a land of progress, wealth, and unstoppable development, Balangero also stated that 'anyone who has read the stories of explorations of the relatively uncharted parts of Australia will also know of the unspeakable torments suffered by those explorers'. He did point out that it was now difficult to get lost in the more populated regions on the coast or along rivers, far from deserts and forests.[149]

Yet, he was lost three times in two years in the woods in northern Queensland, where he worked as a missionary and where there were only two small mining towns, Ravenswood and Charles Towers. Looking for shortcuts along the path he was following, he had experienced the sensation of 'fear and fatigue that grow greater and greater as time passes and hope diminishes'. On the third occasion, Balangero wandered aimlessly in the woods on horseback for a whole day until nightfall. He wrote that it was a 'fear filled night, fortunately with some light from a beautiful moon, which seemed to send me comfort from heaven'. In that circumstance, he was saved by the horse. After he had dismounted and removed the bit to allow the animal to graze, it made its way to the cattle station where it had been raised, about fifteen miles from Ravenswood.[150]

Despite the reference to the moon that shone over the pristine Australian forests, the experience was clearly not pleasant and cannot be compared to a happy experience of nature-based tourism. The three episodes in which

Balangero got lost, and especially the third, illustrate a continuing danger in nature tourism today if undertaken without guides or adequate knowledge of the area being visited. In May 2021, tourists from Victoria went missing in northern Queensland due to bad weather and had to spend the night in the bush before being rescued the next morning. Between March and February 2014, a German tourist was lost for almost three weeks in central-west Queensland after leaving the main roads between the towns of Windorah and Jundah and setting out cross-country.[151]

For Balangero, the experience of having lost his bearings in the forests of Queensland must have left a deep impression: the frontispiece of his book featured a drawing of the missionary on horseback lost in the trees (looking undaunted in a clearly staged pose) opposite the author's portrait on the endpaper.

We close Chapter 1 by underscoring how the case of Balangero and the experiences he had in Australia and Ceylon linked to his role as missionary offer an interesting perspective on the dynamics of nature-based tourism today. The sensitivity and attention to society and the world around him intrinsic to his role allowed him to note information and issues relating to

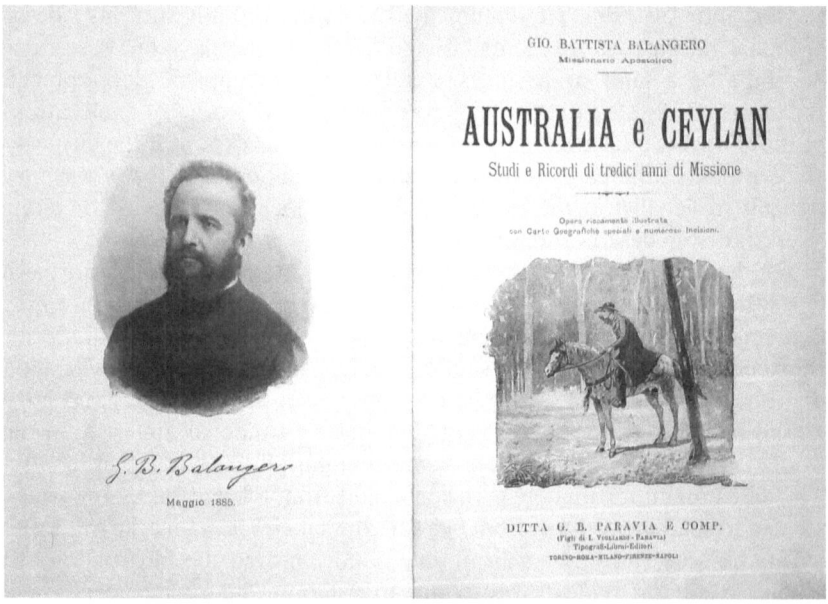

Figure 1.5 Portrait of Giovanni Battista Balangero and representation of him lost in the forests of Queensland. Giovanni Battista Balangero, *Australia e Ceylan: Studi e ricordi di tredici anni di missione* (Turin, 1897), endpaper and frontispiece.

the Matara area in Ceylon and northern Queensland in Australia—issues that are still relevant today regarding the tourism potential of these fascinating places—and communicate them to his compatriots in Italy.

Notes

1 See the definitions of 'Country of Residence' and 'Visitor' offered by the *Glossary of Tourism Terms* in the UNWTO website, https://www.unwto.org/glossary-tourism-terms, accessed on October 17, 2022.

2 For some perspectives on the Grand Tour in its social and culture framework, see Sarah Goldsmith, *Masculinity and Danger on the Eighteenth-Century Grand Tour* (London, 2020); Attilio Brilli and Simonetta Neri, *Le viaggiatrici del Grand Tour: storie, amori, avventure*, (Bologna, 2020); Rosemary Sweet, *Cities and the Grand Tour: The British in Italy, c. 1690–1820* (Cambridge and London, 2012); Jeremy Black, *France and the Grand Tour* (Basingstoke and New York, 2003); Edward Chaney, *The Evolution of the Grand Tour* (London, 1998); Jeremy Black, *The British and the Grand Tour* (Beckenham, 1985). About the connection between science and travel, see Simona Boscani Leoni, Sarah Baumgartner, and Meike Knittel (eds), *Connecting Territories: Exploring People and Nature, 1700–1850* (Leiden and Boston, 2022); Nina Gerassi-Navarro, *Women, Travel, and Science in Nineteenth-Century Americas: The Politics of Observations* (Cham 2017); Ana Simões, Ana Carneiro, Maria Paula Diogo (eds), *Travels of Learning: A Geography of Science in Europe* (Dordrecht, Boston and London, 2003). See also the thematic issue titled *Travelers* of *The Journal of Modern History*, Volume 93, No. 1 (March 2021).

3 For some examples of studies on different kinds of scientific travel, see Pierre-Yves Beaurepaire, *Les Lumières et le Monde : Voyager, Explorer, Collectionner* (Paris, 2019); Katharine Anderson, *Natural history and the scientific voyage*, in Helen Anne Curry, Nicholas Jardine, James Andrew Secord, Emma C. Spary (eds), *Worlds of Natural History* (Cambridge and New York, 2018), pp. 304–318; Paul Smethurst, *Travel Writing and the Natural World, 1748–1840* (Basingstoke and New York, 2012), pp. 16–67; Marco Ciardi (ed), *Esplorazioni e viaggi scientifici nel Settecento* (Milan, 2008), pp. 7–107; Antonio García Belmar and José Ramón Bertomeu Sánchez, *Constructing the Centre from the Periphery*, in Simões, Carneiro, Diogo (eds) *Travels of Learning*, pp. 143–188; Agnese Visconti, *Scienziati e naturalisti dai Balcani a Capo Nord*, in *Europa: Storie di viaggiatori italiani* (Milan, 1988), pp. 200–229.

4 Sophie Brockmann, *The Science of Useful Nature in Central America: Landscapes, Networks and Practical Enlightenment, 1784–1838* (Cambridge and New York, 2020); Daniela Bleichmar, *Botanical conquistadores*, in Curry, Jardine, Secord, Spary (eds), *Worlds of Natural History*, pp. 236–254; John Newsome Crossley, *The Dasmariñases, Early Governors of the Spanish Philippines* (Abingdon and New York, 2016); Daniela Bleichmar, *Visible Empire: Botanical Expeditions and Visual Culture in the Hispanic Enlightenment* (Chicago and London, 2012); Antonio González Bueno and Raúl Rodríguez Nozal, *Plantas americanas para la España ilustrada: génesis, desarrollo y ocaso del proyecto español de expediciones botánicas* (Madrid, 2000).

5 Annika Lindskog, *Constructing and classifying 'The North': Linnaeus and Lapland*, in Cian Duffy (ed), *Romantic Norths: Anglo-Nordic Exchanges, 1770–1842* (Cham, 2017), pp. 75–99; Derek Ratcliffe, *Lapland: A Natural History*, illustrated by Mike Unwin (London, 2005), pp. 3, 25–26; Lisbet Koerner, *Carl Linnaeus in*

his time and place, in Nicholas Jardine, James Andrew Secord, Emma C. Spary (eds), *Cultures of Natural History* (Cambridge, 1996), pp. 145–177.

6 Carl Linnaeus, *Lachesis Lapponica or a Tour in Lapland: Volume I* (London, 1811), pp. 123–124.

7 Linnaeus, *Lachesis Lapponica: Volume II*, pp. 258–259.

8 Linnaeus, *Lachesis Lapponica: Volume I*, p. 202.

9 Ibid., p. 188.

10 Linnaeus, *Lachesis Lapponica: Volume II*, pp. 261–262.

11 On the complex relationship between nature-based tourism, local activities, and sustainability, see Francesco Antonio Anselmi, *Sustainable Tourism Development: Ecotourism and Governance of Glocal Tourism* (Milan, 2020); Øystein Jensen, Frank Lindberg, Damiannah M. Kieti, Bjørn Willy Åmo, James S. Nampushi, *How local traditions and way of living influence tourism: Basecamp explorer in Maasai Mara, Kenya and Svalbard, Norway*, in Joseph S. Chen, Nina K. Prebsen (eds), *Nature Tourism* (Abingdon and New York, 2017), pp. 68–81; Peter Fredman and Liisa Tyrväinen (eds), *Frontiers in Nature-based Tourism: Lessons from Finland, Iceland, Norway and Sweden* (Abingdon and New York, 2011). Also interesting are the perspectives from some decades ago in Tensie Whelan (ed), *Nature Tourism: Managing for the Environment* (Washington D.C., 1991).

12 Linnaeus, *Lachesis Lapponica: Volume II*, pp. 177–179.

13 Linnaeus, *Lachesis Lapponica: Volume I*, p. 154.

14 Paolo Mazzarello, *Spallanzani, Lazzaro*, in *Dizionario Biografico degli Italiani: Volume 93* (Rome, 2018), https://www.treccani.it/enciclopedia/lazzaro-spallanzani_%28Dizionario-Biografico%29/, accessed on December 4, 2022.

15 Lazzaro Spallanzani, *Viaggi alle Due Sicilie e in alcune parti dell'Appennino: Volume I* (Pavia, 1792), pp. xi–xii.

16 Gian Battista Vai, *Light and shadow: the status of Italian geology around 1807*, in Cherry L. E. Lewis, Simon J. Knell (eds), *The Making of the Geological Society of London* (London, 2009), pp. 179–202; Ciardi (ed), *Esplorazioni e viaggi scientifici nel Settecento*, pp. 94–96, 296–377; Paolo Mazzarello, *Costantinopoli 1786: la congiura e la beffa. L'intrigo Spallanzani* (Turin, 2004); Nicoletta Morello, *Lazzaro Spallanzani geopaleontologo dall'origine delle sorgenti alla vulcanologia*, in Giuseppe Montalenti and Paolo Rossi (eds), *Lazzaro Spallanzani e la biologia del Settecento. Teorie, esperimenti, istituzioni scientifiche* (Florence, 1982), pp. 271–281. About the debate between Neptunism and Plutonism see the brief and clear reconstruction in Michael Leddra, *Time Matters: Geology's Legacy to Scientific Thought* (Oxford, Chichester and Hoboken, 2010), pp. 81–93. See also: Jennifer Trusted, *Beliefs and Biology: Theories of Life and Living* (Basingstoke and New York, 2003, 2nd edition), pp. 90–92; Lamberto Laureti, *Italian contributions during the time of Werner relating to plutonism and neptunism – the works of Esprit-Benoit Nicolis de Robilant and Scipione Breislak*, in Helmuth Albrecht and Roland Ladwig (eds), *Abraham Gottlob Werner and the Foundation of the Geological Sciences. Selected Papers of the International Werner Symposium in Freiberg, 19th–24th September 1999* (Freiberg, 2003), pp. 179–187.

17 Spallanzani, *Viaggi alle Due Sicilie: Volume I*, p. 2. For an interesting study on travelers headed to Mount Vesuvius and on their interaction with local people see John Brewer, 'Visiting Vesuvius: Guides, Local Knowledge, Sublime Tourism, and Science, 1760–1890', *The Journal of Modern History*, Volume 93, No. 1 (March 2021), pp. 1–33.

18 Spallanzani, *Viaggi alle Due Sicilie: Volume I*, pp. 3–5.

19 Brewer, 'Visiting Vesuvius'; Goldsmith, *Masculinity and Danger*, pp. 142–143.

20 Spallanzani, *Viaggi alle Due Sicilie: Volume I*, pp. 216–217.

21 Spallanzani, *Viaggi alle Due Sicilie: Volume I*, pp. 219–224, 228–229.

22 Jussi Välimaa, *A History of Finnish Higher Education from the Middle Ages to the 21st Century* (Cham, 2019), pp. 90–95; Peter O. Wacker, *Swedish Settlement in New Jersey Before 1800*, in Carol E. Hoffecker, Richard Waldron, Lorraine E. Williams, Barbara E. Benson (eds), *New Sweden in America* (Newark and London, 1995), pp. 215–248, in particular pp. 224–225.

23 Adolph B. Benson (ed), *Peter Kalm's Travels in North America: The English Version of 1770: Volume 1* (New York, 1966), p. 14 and note 2.

24 Benson (ed.), *Peter Kalm's Travels in North America: The English Version of 1770: Volume 2* (New York, 1966), p. 703.

25 Henry J. Noltie, *The Botanical Collections of Colonel and Mrs Walker: Ceylon, 1830–1838* (Edinburgh, 2013). See also Henry J. Noltie, *Robert Wight and his European botanical collaborators*, in Vinita Damodaran, Anna Winterbottom, Alan Lester (eds), *The East India Company and the Natural World* (London, 2014), pp. 58–79.

26 Anna Maria Walker, 'Journal of an Ascent to the Summit of Adam's Peak, Ceylon', *Companion to the Botanical Magazine*, Volume 1 (1835), pp. 3–14, in particular pp. 6, 8.

27 Ibid., p. 11.

28 Ibid., pp. 4–5.

29 Ibid., p. 8.

30 Apart from Darwin's autobiography, published posthumously in the collection *The Life and Letters of Charles Darwin, Including an Autobiographical Chapter, Edited by His Son, Francis Darwin. In Three Volumes* (London, 1887), for some biographical information refer to: Felix Sprang, *Charles Darwin, The Voyage of the Beagle (1839)*, in Barbara Shaff (ed), *Handbook of British Travel Writing* (Berlin and Boston, 2020), pp. 373–395; Dorismel Díaz, 'Charles Darwin and the Representation of Black Communities in His Travel Narrative', *Hallazgos*, Volume 12, No. 23 (January–June 2015), pp. 231–249; Janet Browne, 'Making Darwin: Biography and the Changing Representations of Charles Darwin', *The Journal of Interdisciplinary History*, Volume 40, No. 3 (Winter 2010), pp. 347–373; Luciana L. Martins, 'A Naturalist's Vision of the Tropics: Charles Darwin and the Brazilian Landscape', *Singapore Journal of Tropical Geography*, Volume 21, No. 1 (March 2000), pp. 19–33.

31 Charles Darwin, *Journal and remarks, 1832–1836*, in *Narrative of the Surveying Voyages of His Majesty's Ships Adventure and Beagle, between the Years 1826 and 1836, Describing their Examination of the Southern Shores of South America and the Beagle's Circumnavigation of the Globe: Volume III* (London, 1839), p. 343.

32 On the relationship between mountains and tourism (sometimes extreme tourism) thorough history and nowadays refer to: Michal Apollo and Viacheslav Andreychouk, *Mountaineering, Adventure Tourism and Local Communities: Social, Environmental and Economics Interactions* (Cheltenham and Northampton MA, 2022); Ghazali Musa, James Higham, Anna Thompson-Carr (eds), *Mountaineering Tourism* (Abingdon and New York, 2015).

33 Darwin, *Journal and Remarks*, pp. 485–486.

34 Ibid., pp. 486–487.

35 Ibid., p. 487.

36 Darwin, *Journal and Remarks*, p. 589.

37 For instance, see Martino Lorenzo Fagnani, *The Development of Agricultural Science in Northern Italy in the Late Eighteenth and Early Nineteenth Century*

(Cham, 2023); Rossano Pazzagli, *Il sapere dell'agricoltura. Istruzione, cultura, economia nell'Italia dell'Ottocento* (Milan, 2008); Lluis Argemí i d'Abadal and Ernest Lluch (eds), *Agronomía y fisiocracia en España (1750–1820)* (Valencia, 1985).

38 Luca Ciancio, *Autopsie della Terra. Illuminismo e geologia in Alberto Fortis (1741–1803)* (Florence, 1995). See also: Maša Surić, Robert Lončarić, Anica Čuka, Josip Faričić, 'Geological issues in Alberto Fortis' *Viaggio in Dalmazia* (1774) – Excursions géologiques extraites du *Viaggio in Dalmazia* d'Alberto Fortis', *Comptes Rendus Geoscience*, Volume 339, No. 9 (August 2007), pp. 640–650; Gilberto Pizzamiglio, *Introduzione*, in Alberto Fortis, *Viaggio in Dalmazia*, Eva Viani (ed) (Venice, 1987), pp. ix–xxx.

39 Alberto Fortis, *Travels into Dalmatia [...] Observations on the Island of Cherso and Osero. Translated from the Italian under the Author's Inspection* (London, 1778).

40 Ibid., p. 211.

41 Larry Wolff, *Venice and the Slaves: The Discovery of Dalmatia in the Age of Enlightenment* (Stanford, 2001); Pizzamiglio, *Introduzione*, pp. xx–xxi. For an interesting perspective of the representations of Eastern Europe from the Renaissance to the twentieth century (Alberto Fortis's included), see Katarzyna Murawska-Muthesius, *Imaging and Mapping Eastern Europe: Sarmatia Europea to Post-Communist Bloc* (New York and Abingdon, 2021).

42 Peter Vodopivec, *Illyrian Provinces from a Slovene Perspective: Myth and Reality*, and Marko Trogrlić and Josip Vrandečić, *French Rule in Dalmatia, 1806–1814: Globalizing a Local Geopolitics*, both in Ute Planert (ed), *Napoleon's Empire: European Politics in Global Perspective* (Basingstoke and New York, 2016), respectively pp. 252–263 and pp. 264–276; Alberto Becherelli, *La politica adriatica e le Province Illiriche*, in Giovanna Motta (ed), *L'imperatore dei francesi e l'Europa napoleonica* (Rome, 2014), pp. 185–194, in particular pp. 188–191; Ivana Pederzani, *I Dandolo. Dall'Italia dei Lumi al Risorgimento* (Milan, 2014), pp. 106–108.

43 Fortis, *Travels into Dalmatia*, p. 137.

44 Ibid., pp. 137, 286.

45 Ibid., pp. 206–207.

46 In a comparative way, on the relation between environment, sustainability, and fishing-related activities, refer to these case studies: Monica Palladino, Carlo Cafiero, Claudio Marcianò, *The role of social relations in promoting effective policies to support diversification within a fishing community in Southern Italy*, in Francesco Calabrò, Lucia Della Spina, Carmelina Bevilacqua (eds), *New Metropolitan Perspectives: Local Knowledge and Innovation Dynamics towards Territory Attractiveness through the Implementation of Horizon/E2020/Agenda2030: Volume 2* (Cham, 2019), pp. 124–133; Vahdet Ünal, Irmak Ertör, Pinar Ertör-Akyazi, Sezgin Tunca, *Making pescatourism just for small-scale fisheries: the case of Turkey and lessons for others*, in Svein Jentoft, Ratana Chuenpagdee, Alicia Bugeja Said, Moenieba Isaacs (eds), *Blue Justice: Small-Scale Fisheries in a Sustainable Ocean Economy* (Cham, 2022), pp. 315–333.

47 About Cavanilles, his scientific training, and his work at the Madrid Botanical Garden refer to: Real Sociedad Económica de Amigos del País de Valencia (ed), *Antonio José Cavanilles (1745–1804): Segundo centenario de la muerte de un gran botánico* (Valencia, 2004); Antonio González Bueno, *Gómez Ortega, Cavanilles, Zea, tres botánicos de la Ilustración: la ciencia al servicio del poder* (Madrid, 2002); Diana E. Soto Arango, 'Cavanilles y Zea: una amistad político-científica',

Asclepio: Revista de historia de la medicina y de la ciencia, Volume 47, No. 1 (1995), pp. 169–196.

48 For a multifaceted analysis of the *Observaciones*, see the many articles in the thematic issue Joan F. Mateu Bellés and Vicenç M. Rosselló i Verger (eds), *Cuadernos de Geografía de la Universitat de València*, Volume 62 (1997), entirely dedicated to this important work by Cavanilles.

49 Antonio José Cavanilles, *Observaciones sobre la historia natural, geographía, agricultura, población y frutos del Reyno de Valencia: Volume I* (Madrid, 1795), pp. 4–5.

50 Ibid., pp. 59–60.

51 Antonio José Cavanilles, *Observaciones sobre la historia natural, geographía, agricultura, población y frutos del Reyno de Valencia: Volume II* (Madrid, 1797), pp. 84–90.

52 An interesting source of information on Humboldt's life is his own autobiography, titled *The Life, Travels and Books of Alexander von Humboldt with an Introduction by Bayard Taylor* (New York, 1859). For some of the most in-depth studies, see Gregor C. Falk, Manfred R. Strecker, Simon Schneider (eds), *Alexander von Humboldt: Multiperspective Approaches* (Cham, 2022); Nicolaas A. Rupke, *Alexander von Humboldt: a Metabiography* (Frankfurt am Main, 2005). Equally interesting is the nonfiction book Andrea Wulf, *The Invention of Nature: the Adventures of Alexander von Humboldt, the Lost Hero of Science* (New York, 2015).

53 Alexander von Humboldt, *Essai sur la géographie des plantes accompagné d'un tableau physique der régions équinoxiales fondé sur des mesures exécutées [...] pendant les années 1799, 1800, 1801, 1802 et 1803 par A. de Humboldt et A. Bonpland* (Paris, 1805), p. 28.

54 Humboldt, *Essai*, pp. 30–31.

55 Darwin, *Journal and Remarks*, p. 590.

56 Alexander von Humboldt, *Voyage aux régions équinoxiales du nouveau continent, fait en 1799, 1800, 1801, 1802, 1803 et 1804: Volume I* (Paris, 1816), pp. 184–185. About the cultivation of taro and its use as food in different areas of the world, see Kenneth F. Kiple, *A Movable Feast: Ten Millennia of Food Globalization* (Cambridge and New York, 2007), pp. 38–39; W. H. R. Rivers, 'Irrigation and the Cultivation of Taro', *Memoirs and Proceedings of the Manchester Literary and Philosophical Society*, Volume 60 (1915–1916), pp. xliv–xlv. For the cultivation of taro in the Canary Islands, see also Marcelino J. del Arco Aguilar and Octavio Rodríguez Delgado, *Vegetation of the Canary Islands* (Cham, 2018), pp. 104–105.

57 Alexander von Humboldt, *Voyage aux régions équinoxiales du nouveau continent, fait en 1799, 1800, 1801, 1802, 1803 et 1804: Volume V* (Paris, 1820), pp. 128–130.

58 Alexander von Humboldt, *Researches Concerning the Institutions and Monuments of the Ancient Inhabitants of America with Descriptions and Views of Some of the Most Striking Scenes in the Cordilleras! Written in French by Alexander de Humboldt and Translated into English by Helen Maria Williams: Volume I* (London, 1814), pp. 220, 349–350.

59 Ibid., pp. 75–76; volume II, pp. 72–75.

60 Maurizia Alippi Cappelletti, *Giglioli, Enrico Hillyer*, and Guido Cimino, *De Filippi, Filippo*, both in *Dizionario Biografico degli Italiani*, respectively in volume 54 (Rome, 2000), https://www.treccani.it/enciclopedia/enrico-hillyer-giglioli_%28Dizionario-Biografico%29/, and volume 33 (Rome, 1987), https://www.treccani.it/enciclopedia/filippo-de-filippi_(Dizionario-Biografico)/, both accessed on December 4, 2022.

61   Enrico Hillyer Giglioli, *Viaggio intorno al globo della r. pirocorvetta italiana Magenta negli anni 1865–66–67–68* (Milan, 1875), pp. 355–357.

62   Ibid., pp. 347–348, 350, 353.

63   Regarding the particular category of cemetery tourism (thanatourism) and its representations, see Sharon Halevi, 'In Sunshine and in Shadow: Adolescent Girls and Thanatourism in the Early American Republic', *Journal of Tourism History*, Volume 12, No. 1 (2020), pp. 71–85; Rodanthi Tzanelli, *Thanatourism and Cinematic Representations of Risk: Screening the End of Tourism* (Abingdon and New York, 2016). Referring to a wider framework is John Lennon and Malcolm Foley, *Dark Tourism: the Attraction of Death and Disaster* (London, 2000).

64   Giglioli, *Viaggio intorno al globo*, p. 60.

65   Ibid., p. 358.

66   Ibid., p. 534.

67   Tilar J. Mazzeo, *Volney, Constantin François de Chasseboeuf, Comte de*, in *Encyclopedia of the Romantic Era 1760–1850* (New York and London, 2004), pp. 1195–1196; Justin Stagl, *A History of Curiosity: the Theory of Travel 1550–1800* (Abingdon, 2005, 2nd edition), pp. 269–296.

68   What was known as Syria at the time roughly encompassed modern-day Palestine, Syria, Jordan, Lebanon, and Israel.

69   Constantin-François Volney, *Voyage en Syrie et en Egypte, pendant les années 1783, 1784 et 1785: Volume I* (Paris, 1787, 2nd edition), p. 237. Regarding Volney's tribulations in Corsica see Joshua Meeks, *France, Britain, and the Struggle for the Revolutionary Western Mediterranean: War, Culture and Society, 1750–1850* (Cham, 2017), pp. 60–62; André Fazi, 'Volney et la Corse', *Bulletin de la Société des Sciences Historiques et Naturelles de la Corse*, No. 718–719 (2007), pp. 27–95.

70   Volney, *Voyage en Syrie et en Egypte: Volume I*, pp. 237–238.

71   Ibid., pp. 16–18; Giovanni Battista Brocchi, *Giornale delle osservazioni fatte ne' viaggi in Egitto, nella Siria e nella Nubia: Volume I* (Bassano del Grappa, 1841), pp. 144–145, 263–265, 329, 331–339.

72   Volney, *Voyage en Syrie et en Egypte: Volume II*, pp. 156–158. Volney visited Tripoli in early 1784 and thus did not directly experience the summer humidity, probably filling in gaps in his information from the available bibliography and conversations with locals. Regarding the chronology of Volney's travels and the seasons in which he made his various visits, see Jean Gaulmier, 'Note sur l'itinéraire de Volney en Égypte et en Syrie', *Bulletin d'Études Orientales*, Volume 13 (1949–1951), pp. 45–50.

73   Constantin-François Volney, *Tableau du climat et du sol des Etats-Unis d'Amérique: Volume I* (Paris, 1803), p. 7.

74   Ibid., pp. 7–8.

75   Ibid., pp. 106–107; the description of Niagara Falls continues to c. p. 124. See also Nigel Leask, *Curiosity and the Aesthetics of Travel-Writing, 1770–1840* (Oxford and New York, 2002), in which the author, referring to Volney's experience in Egypt, highlights how his book 'is deeply paradoxical, sceptical of the efficacy or utility of "personal narrative" as against the purely statistical and topographical account' (p. 112). We can actually foresee a similar perspective also in Volney's description of Niagara Falls and its surroundings.

76   Volney, *Tableau du climat: Volume I*, p. 118.

77   Ibid., pp. 9–12.

78   For some biographical information and his intellectual profile refer to: James T. Kloppenberg, *Toward Democracy: the Struggle for Self-Rule in European and*

American Thought (Oxford and New York, 2016), pp. 452–453; Ed Cohen, *A Body Worth Defending: Immunity, Biopolitics, and the Apotheosis of the Modern Body* (Durham and London, 2009), pp. 133–135; Paul R. Hanson, *Historical Dictionary of the French Revolution* (Lanham, Toronto and Oxford, 2004), pp. 183–184. About his travel to North America, Rochefoucauld writes in the *avertissement* of his work that when he started to write the journal he had in mind to entrust it only to his friends, but some of them advised him to have it published (François de La Rochefoucauld-Liancourt, *Voyage dans les États-Unis d'Amérique fait en 1795, 1796 et 1797: Volume I* [Paris, Year VII], p. v. The introduction written by Henry Neuman for the English version states: 'The Duke … has made a journey for philosophical and commercial observation throughout a great part of North America, and has communicated the substance of his observations to the World, in the valuable Narrative which is here presented to the British public' (François de La Rochefoucauld-Liancourt, *Travels through the United States of North America and the Country of the Iroquois and Upper Canada in the Years 1795, 1796, and 1797; with an Authentic Account of Lower Canada: Volume I* [London, 1799], p. iii). And again: 'It is, amidst all this, impossible not to admire this amiable nobleman, for labouring to divert the taedium of his exile, by enquiries of a tendency so beneficial' (p. xii). These passages clearly show that Rochefoucauld's travels neither were aimed at scientific research nor had diplomatic or business interest.

79 Ibid., pp. 388–389.

80 Ibid., pp. 390–393.

81 François de La Rochefoucauld-Liancourt, *Travels through the United States of North America and the Country of the Iroquois and Upper Canada in the Years 1795, 1796, and 1797; with an Authentic Account of Lower Canada: Volume III* (London, 1800, 2nd edition), pp. 214–215.

82 For a general description of the travels of Mary Wollstonecraft Shelley, see Miranda Seymour, *Mary Shelley* (New York, 2000), pp. 103–113, 146–164, 203–321, 475–496.

83 Paula R. Feldaman and Diana Scott-Kilvert (eds), *The Journals of Mary Shelley 1814–1844: Volume 1* (Oxford, 1987), pp. 116–117. See also Seymour, *Mary Shelley*, pp. 158–159; Jean M. Grove, *Little Ice Ages: Ancient and Modern: Volume I* (London and New York, 2004, 2nd edition), pp. 103–122; Eric. G. Wilson, *The Spiritual History of Ice: Romanticism, Science, and the Imagination* (New York and Basingstoke, 2003), pp. 90, 104–105, 125–128.

84 Feldaman and Scott-Kilvert (eds), *The Journals of Mary Shelley: Volume 1*; for the episode with the Dutch, see p. 116, for the Cascade see p. 114.

85 Mary Wollstonecraft Shelley, *Rambles in Germany and Italy in 1840, 1842, and 1843: Volume I* (London, 1844), pp. 16–17.

86 Ibid., pp. 207–208.

87 Percy Bysshe Shelley and Mary Wollstonecraft Shelley, *Notes to the Complete Poetical Works of Percy Bysshe Shelley* (London, 1839), p. 191.

88 Wollstonecraft Shelley, *Rambles: Volume I*, p. 53.

89 Ibid., pp. 200–201. The UNESCO infosheets on the beech forests in the Carpathians, in Germany, and in other parts of Europe may be found at https://worldheritagegermany.com/germanys-ancient-beech-forests/, accessed on October 19, 2022, and http://whc.unesco.org/en/list/1133, accessed on October 19, 2022.

90 Wollstonecraft Shelley, *Rambles: Volume I*, p. 60.

91 Wollstonecraft Shelley, *Rambles: Volume II*, pp. 267–277.

92 Seymour, *Mary Shelley*, pp. 151–164.

93 Wollstonecraft Shelley, *Rambles: Volume I*, pp. 45–46.

94 Wollstonecraft Shelley, *Rambles: Volume II*, p. 20.

95 Michelle Payne, *Marianne North: a Very Intrepid Painter* (Chicago and Kew, 2016, 2nd edition); Suzanne Le-May Sheffield, *Revealing New Worlds: Three Victorian Women Naturalists* (London and New York, 2001), pp. 75–138. See also the *Introduction* written by Susan Morgan in her own edition of North's *Recollections of a Happy Life: Volume I*, published by the University Press of Virginia (Charlottesville and London, 1993), pp. xii–xl.

96 For the editorial life of the *Recollections*, see Brenda E. Moon, 'Marianne North's *Recollections of a happy life*: how They Came to be Written and Published', *Journal of the Society for the Bibliography of Natural History*, Volume 8, No. 4 (1978), pp. 497–505. See also the preface in Marianne North, *Recollections of a Happy Life: Volume I*, Janet Catherine North as Mrs. John Addington Symonds (ed) (New York and London, 1892), pp. v–vi, where Janet writes: 'My sister was not botanist in the technical sense of the term: her feeling for plants in their beautiful living personality was more like that which we all have for humans friends'. For this reason, we have not included Marianne North in the section on science travel. For an interesting and new-perspective study titled *'Escaping Gender': the Neutral Voice in Marianne North's Recollection of a Happy Life*, see Patricia Murphy, *In Science's Shadow: Literary Constructions of Late Victorian Women* (Columbia and London, 2006), pp. 140–175.

97 Ibid., pp. 51–52.

98 See the abundant offering of cruises, tours, restaurants, historical sites, museums, minigolf courses, and movie theatres at https://www.lakewinnipesaukee.net, accessed on October 19, 2022.

99 North, *Recollections: Volume I*, pp. 51–52.

100 Ibid.

101 North, *Recollections: Volume II*, pp. 217, 280.

102 Ibid., pp. 236–239.

103 Ibid., pp. 284, 308–309.

104 Ibid., pp. 285–287

105 Ibid., pp. 287, 301–302.

106 See De Riseis's personal dossier in the website of the Italian Senate: https://notes9.senato.it/web/senregno.nsf/c1544f301fd4af96c125785d00598476/7439e0ab914cdaab4125646f005b0c8e?OpenDocument, accessed on October 19, 2022. See also Pierfrancesco Fedi, 'Alcune note su Giovanni De Riseis', *Rivista degli studi orientali*, Volume 78, No. 3–4 (2005), pp. 237–259.

107 Giovanni De Riseis, *Il Giappone moderno* (Milan, 1896), introduction page.

108 Ibid., pp. 2–3.

109 Ibid., pp. 6–7, 24–26.

110 Ibid., pp. 52–56.

111 Ibid., pp. 328–335, 338.

112 On Japan's economic development in the Meiji era, see Takatoshi Ito and Takeo Hoshi, *The Japanese Economy* (Cambridge MA and London, 2020, 2nd edition), pp. 10–23; Kenichi Ohno, *The History of Japanese Economic Development: Origins of Private Dynamism and Policy Competence* (Abingdon and New York, 2017), pp. 35–81.

113 Janet Browne, *Missionaries and the Human Mind: Charles Darwin and Robert Fitzroy*, in Roy MacLeod and Philip F. Rehbock (eds), *Darwin's Laboratory: Evolutionary Theory and Natural History in the Pacific* (Honolulu, 1994),

pp. 263–282; Richard Darwin Keynes (ed), *The Beagle Record: Selections from the Original Pictorial Records and Written Accounts of the Voyage of H.M.S. Beagle* (Cambridge and New York, 1979), pp. 1–10; Howard E. L. Mellersh, *FitzRoy of the Beagle* (London, 1968).

114 Payne, *Marianne North: a Very Intrepid Painter*, pp. 8, 13; Le-May Sheffield, *Revealing New Worlds*, pp. 83–106.

115 About Rochon's scientific profile and the context of his travels, see Thomas J. Anderson, *Reassembling the Strange: Naturalists, Missionaries, and the Environment of Nineteenth-Century Madagascar* (Lanham, Boulder, New York and London, 2018), pp. 23–29; Brian Gee, *Francis Watkins and the Dollond Telescope Patent Controversy*, Anita McConnell and A.D. Morrison-Low (eds) (Farnham and Burlington VT, 2014), pp. 231–235; William McAteer, *Rivals in Eden: a History of the French Settlement and British Conquest of the Seychelles Islands (1742–1818)* (Lewes, 1991); Donald McDonald and Leslie B. Hunt, *A History of Platinum and its Allied Metals* (London, 1982), pp. 87–88.

116 Alexis-Marie de Rochon, *A Voyage to Madagascar and the East Indies Translated from the French* (London, 1792), pp. xlv–xlvi, 250–251. About Poivre and the botanical garden at Monplaisir, see Jean-Baptiste Fressoz and Fabien Locher, *Les révoltes du ciel: une histoire du changement climatique XVe-XXe siècle* (Paris, 2020), pp. 66–77; Anderson, *Reassembling the Strange*, pp. 24–26; Jean Paul Morel, *Monplaisir, un jardin bien nommé. Deuxième partie : 1767–1772*, http://www.pierre-poivre.fr/Monplaisir-P2.pdf, accessed on October 19, 2022. On the global transfer of plant species and the role of botanical gardens in economic development during the eighteenth and nineteenth centuries, see Lance C. Turner, 'Botanizing in the Borderlands: the Limits of Scientific Indigeneity in Late Colonial New Spain', *Colonial Latin American Review*, Volume 30, No. 1 (2021), pp. 109–136; Barry L. Stiefel, *Maple: The Sugar of Abolitionist Aspirations*, and Christopher Magra, *Chocolate and the Atlantic Economy: Circuits of Trade and Knowledge*, both in Victoria Barnett-Woods (ed), *Cultural Economies of the Atlantic World: Objects and Capital in the Transatlantic Imagination* (New York and Abingdon, 2020), respectively pp. 147–172 and pp. 173–190; Londa Schiebinger and Claudia Swan, *Colonial Botany: Science, Commerce, and Politics in the Early Modern World* (Philadelphia, 2005); Emma C. Spary, *Utopia's Garden: French Natural History from Old Regime to Revolution* (Chicago and London, 2000); Lucile H. Brockway, *Science and Colonial Expansion: The Role of the British Royal Botanic Gardens* (New York and London, 1979).

117 Rochon, *A Voyage to Madagascar*, pp. 7–9.

118 Ibid., pp. 5–6.

119 Ibid., pp. 349–350, 351–352.

120 Ibid., pp. 5, 252–253, 338, 347.

121 Dario Manfredi, *Malaspina, Alessandro*, in *Dizionario Biografico degli Italiani: Volume 67* (Rome, 2006), https://www.treccani.it/enciclopedia/alessandro-malaspina_%28Dizionario-Biografico%29/, accessed on December 4, 2022; Donald D. Cutter, *Introduction*, in Andrew David, Felipe Fernandez-Armesto, Carlos Novi, Glyndwr Williams (eds), *The Malaspina Expedition 1789–1794: The Journal of the Voyage by Alessandro Malaspina: Volume 1* (London and Madrid, 2001), pp. xxix–lxxvii.

122 All quotes from *The Malaspina Expedition 1789–1794: Volume 1*, p. 209.

123 Jan M. Tapia-Guerra, Ariadna Mecho, Erin E. Easton, María de los Ángeles Gallardo, Matthias Gorny, Javier Sellanes, 'First Description of Deep Benthic Habitats and Communities of Oceanic Islands and Seamounts of the Nazca

Desventuradas Marine Park, Chile', *Scientific Reports*, Volume 11, No. 6209 (2021), https://doi.org/10.1038/s41598-021-85516-8, accessed on October 19, 2022; National Geographic and Oceana (eds), *Islas Desventuradas: biodiversidad marina y propuesta de conservación. Informe de la expedición 'Pristine Seas' National Geographic Society / Oceana, febrero del 2013*, https://media.nationalgeographic.org/assets/file/PristineSeasDesventuradasScientificReport.pdf?_gl=1*vjdwpf*_ga*MTg4NzQzODEyOC4xNjU3NzE4NzEz*_ga_JRRK GYJRKE*MTY1NzcxODcxMy4xLjEuMTY1NzcxODk0OS4w, accessed on October 19, 2022.

124 *The Malaspina Expedition 1789–1794: Volume 2*, p. 142.
125 Ibid., p. 152.
126 Ibid., p. 159.
127 Brian Vander Naald, 'Examining Tourist Preferences to Slow Glacier Loss: Evidence from Alaska', *Tourism Recreation Research*, Volume 45, No. 1 (2020), pp. 107–117; Lee K. Cerveny, *Sociocultural Effects of Tourism in Hoonah, Alaska* (Washington D.C., 2007); Lee K. Cerveny, *Tourism and Its Effects on Southeast Alaska Communities and Resources: Case Studies from Haines, Craig, and Hoonah, Alaska* (Washington D.C., 2005).
128 All quotes from *The Malaspina Expedition 1789–1794: Volume 2*, p. 287.
129 Valerio Giacomini, *Brocchi, Giovanni Battista*, in *Dizionario Biografico degli Italiani: Volume 14* (Roma, 1972), https://www.treccani.it/enciclopedia/giovanni-battista-brocchi_%28Dizionario-Biografico%29/, accessed on December 4, 2022; Brocchi, *Giornale delle osservazioni: Volume I*, pp. xv–xviii. About the Egypt of Mehmet Ali, see Atul Kohli, *Imperialism and the Developing World: How Britain and the United States Shaped the Global Periphery* (Oxford and New York, 2020), pp. 102–104; Khaled Fahmy, *Mehmed Ali: From Ottoman Governor to Ruler of Egypt* (London, 2009); Khaled Fahmy, *All the Pasha's Men: Mehmed Ali, His Army and the Making of Modern Egypt* (Cambridge and New York, 1997); Afaf Lutfi Al-Sayyid Marsot, *Egypt in the Reign of Muhammad Ali* (Cambridge and New York, 1984).
130 Brocchi, *Giornale Delle Osservazioni: Volume I*, pp. 145–146.
131 Giovanni Battista Brocchi, *Giornale delle osservazioni fatte ne' viaggi in Egitto, nella Siria e nella Nubia: Volume IV* (Bassano del Grappa, 1843), pp. 196–197.
132 Giovanni Battista Brocchi, *Giornale delle osservazioni fatte ne' viaggi in Egitto, nella Siria e nella Nubia: Volume III* (Bassano del Grappa, 1842), pp. 83–84, 117–25.
133 Ibid., p. 341. As regards the classical tones characterizing his technical and scientific discourse see Martino Lorenzo Fagnani, 'Travels and Representations at the Core of Western Agricultural Science: Discovering Rural Societies in Spain, Italy and Lebanon, Late Eighteenth and Early Nineteenth Centuries', *Continuity and Change*, Volume 37, No. 3 (December 2022), pp. 313–334.
134 Brocchi, *Giornale delle osservazioni: Volume IV*, pp. 95–98. Brocchi writes about a 'dura bread' (*pane di dura*), which can be identified with bread-type foods made with sorghum: Food and Agriculture Organization of the United Nations (ed), *Sorghum and Millets in Human Nutrition* (Rome, 1995), pp. 1–3; Office of International Affairs National Research Council (ed), *Application of Biotechnology to Traditional Fermented Foods: Report of an Ad Hoc Panel of the Board on Science and Technology for International Development* (Washington D.C., 1992), pp. 27–28.
135 Brocchi, *Giornale delle osservazioni: Volume III*, pp. 33–34.
136 *Quran*, Surah 24:39, in Brocchi, *Giornale delle osservazioni: Volume I*, pp. 358–359.

137 Brocchi, *Giornale delle osservazioni: Volume I*, pp. 346–355.

138 For an interdisciplinary overview on the perception and representation of nature by travelers and modern-day tourists, see, for example, these interesting studies: Jon Mathieu, *Divergent perception: deserts and mountains in transition to modernity, seen through Alexander von Humboldt's* Views of Nature, in Boscani Leoni, Baumgartner, Knittel (eds), *Connecting Territories*, pp. 189–209; Ida Ruffolo, *The Perception of Nature in Travel Promotion Texts: A Corpus-based Discourse Analysis* (Bern, 2015); Lesego Senyana Stone, *Perceptions of Nature-based Tourism, Travel Preferences, Promotions and Disparity between Domestic and International Tourists: The Case of Botswana* (PhD thesis, Arizona State University, 2014), https://keep.lib.asu.edu/_flysystem/fedora/c7/124454/Stone_asu_0010E_14487.pdf, accessed on October 21, 2022; Paige West, 'Tourism and Science and Science as Tourism: Environment, Society, Self, and Other in Papua New Guinea', *Current Anthropology*, Volume 49, No. 4 (August 2008), pp. 597–626; Mary Louise, *Imperial Eyes: Travel Writing and Transculturation* (London and New York, 1991).

139 For a wide view of different Christian missionary activities through history, among the scientific literature of the last two decades, see Jenna M. Gibbs (ed), *Global Protestant Missions: Politics, Reform, and Communication, 1730s–1930s* (Abingdon and New York, 2020); Catherine Bal、riaux, *Missionary Strategies in the New World, 1610–1690: An Intellectual History* (New York and Abingdon, 2019); Robert Aleksander Maryks and Jonathan Wright (eds), *Jesuit Survival and Restoration. A Global History, 1773–1900* (Leiden and Boston, 2015), in particular Part 4, Part 5, and Part 6, collecting studies on the dynamics of Jesuit missions in Asia, the Americas, and Africa; Helen May, Baljit Kaur, Larry Prochner, *Empire, Education, and Indigenous Childhoods: Nineteenth-Century Missionary Infant Schools in Three British Colonies* (Farnham and Burlington VT, 2014); Amanda Barry, Joanna Cruickshank, Andrew Brown-May, Patricia Grimshaw (eds), *Evangelists of Empire? Missionaries in Colonial History* (Melbourne, 2008); Brian Stanley (ed), *Christian Missions and the Enlightenment* (Abingdon and New York, 2001).

140 Giovanni Battista Balangero, *Australia e Ceylan: Studi e ricordi di tredici anni di missione* (Turin, 1897), p. vi (the work includes an appendix with Italian translations of some letters Balangero wrote about Palestine); *Lunario genovese compilato dal signor Regina & C. – Guida amministrativa e commerciale di Genova e provincia* (Genoa, 1887), p. 261; *Bollettino ufficiale del Ministero della Pubblica Istruzione*, Volume 3, No. 7–8 (July–August 1877), p. 479; *Il Movimento Cattolico*, October 15, 1885, p. 299; Ministero della Pubblica Istruzione (ed), *Ruoli di anzianità al 1° agosto 1902, edizione provvisoria* (Rome, 1902), p. 309.

141 Balangero, *Australia e Ceylan*, pp. v, 5–6, 181–182.

142 Balangero, *Australia e Ceylan*, pp. v–vi.

143 On the history of globalization, see Julia Zinkina, David Christian, Leonid Grinin, Ilya Ilyin, Alexey Andreev, Ivan Aleshkovski, Sergey Shulgin, and Andrey Korotayev (eds), *A Big History of Globalization: The Emergence of a Global World System* (Cham, 2019); Kevin H. O'Rourke and Jeffrey G. Williamson, *Globalization and History: The Evolution of a Nineteenth-Century Atlantic Economy* (New York, 1999).

144 Balangero, *Australia e Ceylan*, pp. 196–200.

145 Ibid., p. 221.

146 Anslem de Silva, 'Crocodiles: Our Living Dinosaurs (With Notes on The *kimbul kotuwa* or The Crocodile Excluding Enclosures)', *Loris: Journal of the Wildlife*

and *Nature Protection Society of Sri Lanka*, Volume 27, No. 5–6 (December 2016), pp. 22–27.

147 Dinal J. S. Samarasinghe, *The Human-Crocodile Conflict in Nilwala River, Matara (Phase 1)* (Colombo, 2014). For the same issue but from a point of view extended to the entirety of Sri Lanka, see A.A. Thasun Amarasinghe, Majintha B. Madawala, D.M.S. Suranjan Karunarathna, S. Charlie Manolis, Anslem de Silva, Ralf Sommerlad, 'Human-crocodile conflict and conservation implications of saltwater crocodiles *Crocodylus porosus* (Reptilia: Crocodylia: Crococodylidae) in Sri Lanka', *Journal of Threatened Taxa*, Volume 7, No. 5 (April 2015), pp. 7111–7130.

148 Balangero, *Australia e Ceylan*, pp. 221–222. For the contemporary events mentioned earlier, see A. W. Gunawardhana, *Crocodile Kills Man in Nilwala Ganga*, September 5, 2007, http://archives.dailynews.lk/2007/09/05/news16.asp, accessed on October 21, 2022; *Crocodile attacks kills police officer in Matara*, July 2, 2020: https://www.newsfirst.lk/2020/07/02/crocodile-attacks-kills-police-officer-in-matara/, accessed on October 21, 2022.

149 Balangero, *Australia e Ceylan*, p. 176.

150 Ibid., pp. 176–180.

151 Elizabeth Cramsie, *Two Victorian Tourists Found Alive After Getting Lost Bushwalking in North Queensland*, May 22, 2021: https://www.abc.net.au/news/2021-05-22/bushwalkers-victoria-missing-north-queensland-search-mareeba/100157784, accessed on October 21, 2022; *Australia: Lost German Tourist 'Ate Flies' to Survive*, March 7, 2014: https://www.bbc.com/news/world-asia-26463979, accessed on October 21, 2022.

2 Environments

2.1 Introduction

The analysis of environments is very important for understanding tourism in natural and agricultural ecosystems. Each environment, including natural environments, is not in itself immutable but is a diachronic succession of the human-territory relationship. The environment, therefore, is a repository of evidence that tells of changes wrought by time. Moreover, descriptions of landscapes are related to each individual's perceptions and social, cultural, and emotional background. Tourists, therefore and nevertheless, fix snapshots that are evocative of a complex and ever-changing system.[1]

In the eighteenth century, Alexander von Humboldt emphasized that to fully enjoy natural beauty, it was necessary to combine it with science and literature. In his scientific work *Kosmos*, Humboldt described environments as the product of human intuition and reason related to the perception of nature. However, as Carlo Cencini points out, there are conceptual difficulties that arise in having to provide a definition of natural and agricultural ecosystems, which would necessitate an almost impossible separation of the elements of nature from the cultural elements that closely relate to them, making objectivity nearly unachievable.[2]

The environments of natural and agricultural ecosystems can thus possess two different meanings depending on the subjective or objective view in which they are considered. The subjective view is related to the feelings caused by the constituent elements of the environment, while the objective view of the landscape locates its roots in physical geography and the natural sciences and aims to study the constituent elements of the environment. To this end, the science of ecology was born in the nineteenth century, allowing for a precisely codified interpretation of the natural and anthropogenic aspects that make up natural and agricultural ecosystems.[3]

There is a further classification related to the fact that environments can take on different characteristics based on the perceptions and benefits they are able to produce in observers. Tiziano Tempesta discusses how 'instinctive

DOI: 10.4324/9781003230519-3

perception' causes tourists to prefer environments that allow them an easy understanding of the natural and agricultural ecosystems around them, stimulating their curiosity. 'Affective perception', in contrast, develops with age, education, culture, and personal experiences and refers to tourists seeking a manifestation or reflection of their own emotions in a given setting.[4]

Within a natural or agricultural ecosystem, the tourist can assume the dual role of actor and spectator. Stepping out of a daily routine, they can assume the position of observer or a participatory agent in the given context, better understanding the peculiar scientific and cultural characteristics of the environments.[5]

Attention to natural and agricultural ecosystems is then also a highly topical issue: it has aroused growing interest internationally in recent decades as a result of increased collective awareness of environmental issues, the overexploitation of biomass and natural lands, and the impact of agriculture on the environment.[6]

In this sense, the history of tourism, related to the themes of natural and agricultural ecosystems, can represent a new area of study in environmental history, which considers the mutual relationships between humankind and the rest of nature over time. John Robert McNeill in 2003 pointed out the exponential growth in the previous two decades of the international literature on the subject and thus the substantial impossibility of proposing a unified picture of the discipline, having to rely on a few more or less in-depth surveys. The country in which environmental history has become most prominently established is the United States, where the discipline initially developed in the 1970s through the work of a number of historians involved in the environmentalist militancy of that period. As McNeill notes, these scholars came predominantly from the field of 'new western history' and thus sought to reinterpret some of the classic themes of that strand of U.S. historiography from a new perspective. These themes include the concept of wilderness, the colonization of the West, and the rise of the conservationist movement.[7]

Environmental history has witnessed a contemporary institutionalization that has occurred through the birth and spread of university courses (including with chairs in Environmental History) and the founding of specialized scholarly journals. In 1976 the first environmental history journal was founded (initially *Environmental Review*, now *Environmental History*), and the following year, the American Society for Environmental History was established. Australia, Canada, New Zealand, and South Africa have witnessed, on a smaller but nonetheless relevant scale, a similar spread of interest, both in academia and in popular science.[8]

For these reasons, historical tourism analysis related to natural and agricultural ecosystems represents an innovative key for analyzing environmental history issues as well, thanks to emblematic case studies.

2.2 Agricultural Ecosystems

Tourism in agricultural ecosystems is related to the cultural perception of environments, which is an essential element in the ability to correctly interpret a territory, including local agricultural practices. An agricultural ecosystem environment describes the evolution it has undergone over time as a result of the layered accumulation of transformation actions by humans.[9] Obviously by agricultural ecosystems, we mean those areas where agricultural activity has been a significant factor in the local economy and work practices.

Italian physician and naturalist Ciro Pollini (1782–1833) made frequent trips in the 1810s to the Lake Garda area, in northern Italy, and especially to Mount Baldo in order to observe its peculiar flora, typical of the lake ecosystem. After receiving a medical degree from the University of Pavia in 1802, Pollini made the choice to devote himself to botanical studies. He taught botany and agricultural science in Milan and Verona. Pollini applied himself to the study of the flora found in the Verona area and the Lake Garda area, publishing several works on the subject. Over the course of his career, he studied various plant and animal species. Indeed, a species of cochineal that is particularly harmful to olive trees in the Mediterranean basin and that he studied on the trees of Lake Garda was later named in his honor: *Pollinia pollinii* (Costa, 1857).[10]

In his tours, Pollini described these places as 'delightful truly and pleasant above the imagination of those who have not seen them'.[11] In relation to the agricultural ecosystem, he noted how the localities near the lake shore possess agriculture characterized by tree crops, especially plots with mulberry, olive, and grapevine plantings. In addition, as already pointed out, the presence of the lake allowed the cultivation of citrus fruits, especially lemons, oranges, and citrons, to be developed in this specific agricultural ecosystem, whereas they were not common in other areas of the Po River Plain. Pollini also reported olfactory impressions, as he was surprised by the fragrance developed by citrus plants and diffused in the air. On the ecosystem of the lake, Pollini noted that the plants were almost all those found in aquatic and marshy places in the Italian context. However, he wrote, 'various plants of the southern regions sprout from crevices in the cliffs'. But only a few kilometers away, on the hills surrounding the lake but at a higher altitude, there was already a different agricultural ecosystem, where citrus groves gave way to 'robust olive trees, grapevines, and other fruit trees', along with meadows and odoriferous grasses.[12]

The scenery of Lake Garda, and its natural agricultural ecosystems, has also been described by Goethe, who visited the area on his Grand Tour. In his travel diary from the village of Torbole in September 1787, he confesses,

How much do I wish that my friends were with me for a moment to enjoy the prospect, which now lies before my eyes. I might have been in Verona this evening but a magnificent natural phenomenon was in my vicinity—Lake Garda, a splendid spectacle, which I did not want to miss, and now I am nobly rewarded for taking this circuitous route.[13]

Very interesting in relation to the analysis of agricultural ecosystems is the case of clergyman Carlo Amoretti (1741–1816). He came from a wealthy family in the Liguria area, taught ecclesiastical jurisprudence at the University of Parma, and later worked as a librarian and private tutor in Milan. He was also the secretary of the Patriotic Society founded in Milan by Maria Theresa of Austria for the improvement of the agricultural and manufactural sectors and later became a member of the Council of Mines of the Kingdom of Italy under Napoleonic rule.[14]

His interest in travel (in Italy and to Switzerland, France, and Austria), either together with patrons or colleagues or alone, at times on behalf of the institutions for which he worked, is evident in his intense activity as a translator and writer. He was an acquaintance of English agriculturist Arthur Young, whom he accompanied in 1789 on visits to Tuscany and Emilia during his tour of the continent. Generally speaking, Amoretti was a sensitive observer of agricultural ecosystems. He often sketched them briefly but nevertheless conveys his interest in and awareness of their elements, even in relation to geomorphological characteristics.[15]

For instance, in the depiction of his journey between Milan and Nice, he described many different regions. The irrigated Po Valley between Milan and Pavia is portrayed as an example of centuries of human work in organizing agricultural systems. Canals for trade and irrigation and smaller ditches traced out an environment whose main agricultural elements were rice fields and permanent meadows.[16] This area was the object of attention by many, especially those with an interest in the organization and management of agricultural ecosystems.[17] Furthermore, the hills of Piedmont represented an agricultural ecosystem that is still famous worldwide for the production of wines. The hills of Gavi, famous for white wines, were described by Amoretti in the following way: 'It is seen that the interior of those hills is a mixture of rolled, scattered, and broken stones, on which are a few feet of topsoil sufficient to foster and support vigorous vegetation, either grapevines, wheat, or chestnut trees'.[18]

Amoretti also described the agricultural ecosystems of Liguria, where the mild marine climate is favorable to typically Mediterranean crops. It is interesting to note, among other things, that it was in the middle decades of the nineteenth century that the western end of the Ligurian Riviera became a popular international destination for climatic and seaside tourism. This occurred many decades after Amoretti's travels and descriptions thereof, but it was the same mild climate and pleasant setting of the western Ligurian Riviera and its French counterpart that made them particularly appreciated by international elites from Great Britain to Russia.[19]

Dominating the agricultural ecosystems described by Amoretti were olive groves. Starting from Varazze, these groves became more common and more extensive westward, becoming the sole product of some towns, where the fruits were smaller but they gave better oil. In a concise way, Amoretti thus described the cultivation of Taggiasca olives, from which a renowned high-quality oil is still produced. The other widespread crop was citrus fruits. It is interesting to note that within this description of a Ligurian agricultural ecosystem, some areas/terroirs that specialize in different cultivation can be distinguished. Thus, for example, on the Albenga plain, alongside the cultivation of olive trees, there were carob trees and vineyards grown intermixed with figs and other fruit trees, again often citrus.[20]

The agricultural ecosystems change as one moved inland into the first hills. For example, a different kind of cultivation was seen in the area of Vezzi. In addition to the vineyards and fields of grain, extensive plantations of figs, whose dried fruit provided winter food for the local inhabitants, took the place of citrus and olive groves. Western Liguria, starting at the valley of Diano Marina, was thickly cultivated with olive trees, vineyards, citrus, and figs. In Taggia, in addition to olive trees, a grape was grown that yielded a renowned Muscat. San Remo, by comparison, was particularly noted for its abundance of citrus fruits. Indeed, not content with their extensive gardens and fruit orchards, the inhabitants had planted them under the olive trees, which were trained higher to accommodate them. Moreover, the particular climate of the Ligurian Riviera had made Bordighera the northernmost European location where date palms (*Phoenix dactylifera* L.) grew wild, although few of their fruits were able to ripen. Amoretti pointed out that these palm groves also provided additional bounty; Bordighera traders provided much of Italy and Germany with 'fronds of these trees' for the Christian Palm Sunday, celebrated every year on the Sunday before Easter.[21]

Amoretti's peregrinations also allow us to consider other agricultural ecosystems, present instead in the northwestern part of Lombardy, which is characterized by some of the largest lakes in Italy: Lake Maggiore, Lake Como, and a portion of Lake Lugano. Amoretti noted, for example, that in the area of Intra and Pallanza, towns on the shore of Lake Maggiore, trade and manufacturing and craft activities did not undermine agriculture. In particular, the area was famous for vineyards stretching from the lakeshore to the lower hillsides that produced excellent wines.[22]

In the Como area, the shoreline of Lake Como was home to an agricultural ecosystem that allowed the cultivation of plant species that were unusual at that latitude, as was also the case on Lake Garda, as Ciro Pollini witnessed a few years later. Alongside the mulberry groves involved in silk production and the vineyards, olive and citrus trees were also grown on the lakeside. The citrus trees, however, must be covered during the winter as protection from the cold. From Amoretti's description, it is also clear how agricultural ecosystems are transformed over time by both human intervention and climate variations. Amoretti noted how olive cultivation was much more widespread in earlier centuries but that a number of cold winters—especially in 1709—led growers to abandon olive groves in favor of mulberries and silk production. However, he noted that excellent oil had been produced and hoped that such cultivation could regain the importance it had had in the past.[23] In contrast to the Lecco branch of the lake, the area around Como was characterized at that time by mulberry cultivation and silk production. Amoretti observes in this regard that in the Como area 'excellent, and better than on the plain, are the production of cocoons and the quality of the silk'.[24]

It is important to note how innovations in agricultural ecosystems captured the attention of these 'tourists', such as Pollini and Amoretti. For example, moving away from Lake Como in the direction of the village of Tradate, Amoretti described an agricultural ecosystem where wheat, vineyards, and mulberry cultivation were widespread, but he also commented 'that it would be desirable for more grazing lands to be introduced there, and thus more livestock, increasing the functionality and entrepreneurship of farms'. Regarding the nearby lands of Mozzate, Amoretti reported that 'the woods of Count and Cavalier Castiglioni deserve to be visited. The latter brought us from North America, where he was led by the desire to educate himself and to benefit his country, many new trees, which are now numerous in these woods'.[25]

The Mozzate estate was also appreciated by the aforementioned Arthur Young when he visited northern and central Italy, accompanied by Amoretti. As for the wide variety of plants cultivated by the two Castiglioni brothers—Alfonso and Luigi—they were the result both of seeds brought back home by Luigi from his trip to North America in the 1780s and of the network of contacts Alfonso had with foreign botanical institutions. For example, Alfonso made attempts to grow exotic plants from seeds he received from the Madrid Botanical Garden, and Luigi showed some seeds of Seneca snakeroot from North America (*Polygala senega* L.) to the Patriotic Society of Milan and described the medical properties of this plant, successfully urging his colleagues to conduct cultivation experiments.[26] Moreover, Luigi played an important role in fostering contacts and exchanges between the Patriotic Society of Milan and the American Philosophical Society of Philadelphia, led by Benjamin Franklin as president and Benjamin Rush as secretary.[27]

These aspects are part of a phenomenon of globalization of the plant kingdom that began with the Columbian exchanges, which Alfred W. Crosby addressed in his environmental history *The Columbian Exchange*.[28] In the eighteenth and early nineteenth centuries, the study of plant species from the vast territories still controlled by some European countries received a new impetus, with a few major botanical gardens playing a central role—such as the ones in Madrid and Paris—and their collaboration with naturalists from both all over Europe and the colonies.[29]

Like Amoretti, Mary Wollstonecraft Shelley—whose travels at different European latitudes we examined in Chapter 1—showed an interest in the agricultural ecosystems of the Lake Como area. Touring the area in 1840, the author noted how the very characteristic vegetation changed quite visibly as soon as she crossed the Swiss border. She immediately noticed the groves of chestnut trees, one of the typical species of the lake area. Describing the area surrounding her hotel, she wrote about the terraces on the mountainsides and listed the crops and vegetation growing there. In addition to chestnut trees, she mentioned olive trees, vineyards, and corn. Among elegant villas, she noted the terraces covered with grapevines and olive trees. Olive cultivation in the Lake Como area was the object of great interest by the political and scientific authorities in Milan since the 1770s. They promoted the acclimatization of hardy olive varieties in the Lake Como area and provided various incentives to local growers. And indeed, olive groves became a stable element of the Lake Como ecosystem, as Wollstonecraft Shelley observed in 1840.[30]

In every description by Wollstonecraft Shelley, the richness of the vegetation is always highlighted. Even the famous *Murray's Handbooks for Travellers* indicated the Kingdom of Lombardy-Venetia under Austrian rule—covering part of northern Italy from 1815 until the late 1850s and the 1860s—as 'the most generally productive part of Italy' and explains that, as far back as the thirteenth and fourteenth centuries, despite those regions' high population and thus high demand, they managed to export large quantities of grain.[31] This prosperity was still evident in Wollstonecraft Shelley's day.

The Murray guidebook divided the Kingdom of Lombardy-Venetia into different areas according to geographical and climatic conditions and made Lake Como part of the 'Littoral Region, or Riviera'.[32] Locations falling within this area are protected from cold winds by the mountains, mitigated by the lakes, but remain exposed to warm currents from the south. This allows the cultivation of plants such as olives, lemons, and laurel. Mulberry trees and vineyards are also common. These conditions also lended themselves to silkworm breeding, one of the main resources of the area. Wollstonecraft Shelley noted that most of the local women were employed in the silk industry and commented on the presence of a spinning mill near the hotel where she was staying, from which she often saw the girls returning home in the evening.[33] The guidebook also reported that this work was done mostly by young women: 'A woman, seated at a vessel containing hot water, prepares and arranges the cocoons, while a girl turns the wheel on which the silk is wound'.[34]

Silkworm cultivation and silk processing during this period stood out not only for their economic importance to the area but also for the very high quality of the silk. As we read again in the handbook: 'All things considered, Italy ranks higher for her silk than any other nation. She supplies her own extensive manufactures, and exports largely, and her prices fix the universal prices of the article'.[35] Of this large production, one-third came precisely from the Kingdom of Lombardy-Venetia, and in particular from the Como area. Among other activities that created employment, Wollstonecraft Shelley naturally mentioned fishing, along with lake transportation, also characteristic features of the lake ecosystem. She relates that 'there is a good deal of traffic on the lake, which is carried on by flat-bottomed barges, impelled by large heavy sails, or by long oars, which they work by pushing forward'.[36]

A particularly significant case study in relation to agricultural ecosystems is the area of the Po River basin in northern Italy. This area has been a favorite destination for tourism in agricultural ecosystems, and numerous travelers have written comprehensive analyses and descriptions of the environments of these territories. Tourists are attracted to the countryside even more than they are to the cities, with the former characterized by impressive water management systems and related agricultural activities. It is a fertile and prosperous land where centuries of development are evident. Travelogues began to proliferate in the second half of the eighteenth century, especially regarding Lombardy, focusing on these bucolic settings along with descriptions of farms, livestock, waterworks, and cheese production. There was often a scientific or agricultural purpose to the travels, but the emotional appeal of the experience is also clearly communicated.[37]

This personal aspect is evident, for example, in the travelogue of the Austrian agriculturist Johann Burger (1773–1842):

> Having passed the Adda River one comes to the land of irrigation. Vineyards and mulberry trees gradually disappear and nothing but irrigated meadows and fields can be seen.... These meadows, which I now saw for the first time, attracted all my attention; I could not sufficiently admire the luxuriant growth of the grass, the deep green of the trees, the linear layouts of the irrigation canals, the mathematical precision of the leveling and sloping of the banks of the water meadows, the simplicity of making the waters flow and drain.[38]

The context of the agricultural ecosystems presented by Burger deserves to be explored in depth, because he visited numerous territories and analyzed and described numerous farms. Burger was born in Wolfsberg, Carinthia, in today's southern Austria. Like his father, he entered the medical profession after studying in Vienna and Freiburg. He applied himself as an autodidact to the study of chemistry and botany. Beginning in 1804, he decided to

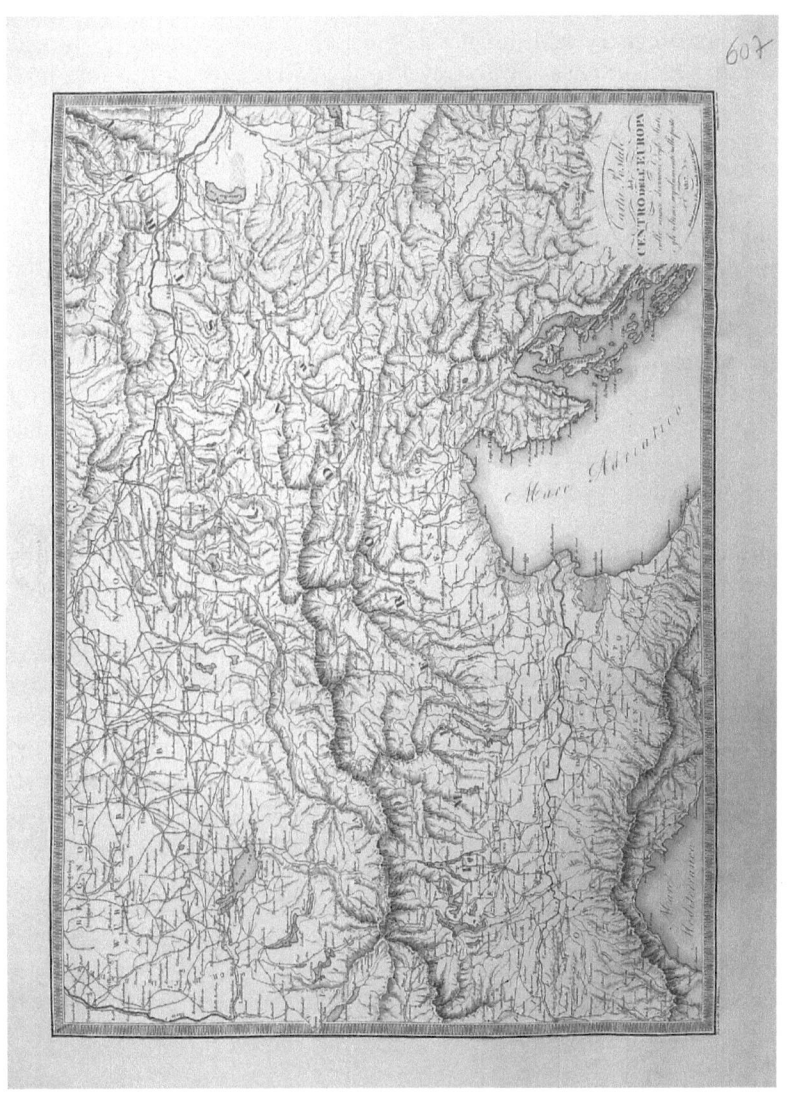

Figure 2.1 The Kingdom of Lombardy-Venetia in the Austrian Empire, 1817. Biblioteca Universitaria di Pavia, Carte geografiche busta F, cartella 9, mappa 607.

devote himself to experimentation in agriculture, in both crop development and the invention of agricultural machinery, also making numerous trips to different countries to learn about their practices and crops. His studies earned him an appointment as a professor of agriculture at Klagenfurt High School. He also had assignments from the government, especially in connection with land registry reforms. It was precisely for inspections of the old cadaster that he had occasion to travel to the Milan area, a trip that produced the text *Reise durch Ober-Italien* (Travels through Northern Italy) in 1831, translated into Italian as *Agricoltura del Lombardo-Veneto* (literally Agriculture of Lombardy-Venetia) in 1843.[39]

The agrarian landscape of the Po Valley, especially in Lombardy, appears to have been shaped over time by human intervention, especially in relation to the use of water, the link between agriculture and cattle breeding, and the wise use of crop rotation. In relation to this last element, the agricultural ecosystems visited by Burger in Lombardy-Venetia are particularly interesting, and he provides an in-depth description, as this was a particularly relevant topic at the European level at the time.[40] Burger observed that the crop rotation system, especially in irrigated agricultural ecosystems, was highly developed, as he was able to observe in the provinces of Milan, Lodi, and Pavia. In the latter, the village of Roncaro featured a seven-year crop rotation cycle: first year, corn; second year, clover; third and fourth year, irrigated meadow; and fifth, sixth, and seventh year, rice. In non-irrigated areas, however, the rotation system adhered to a four-year cycle, with a rotation of wheat, corn, clover, and then one year with oats or millet or flax.[41]

In general, visits to numerous farms in different areas of Lombardy-Venetia allowed Burger to provide a great deal of information on many aspects of the agricultural ecosystems of an area that was being recognized internationally as particularly emblematic of the development of the human–environment relationship over time. Indeed, this is a recurring element in Burger's and many other travelers' descriptions of the region, namely, how irrigation systems and certain agricultural practices are the result of a long history.[42]

2.3 Exploring the Alps

Scientific explorations in the eighteenth and nineteenth centuries developed knowledge of the world's mountain ranges, including the Alps, the Andes, and the Himalayas.

In this section, we analyze the Alps as a case study of tourism in natural ecosystems. The development of mountain tourism was preceded by an enduring negative view of the mountains. Since Roman times, the mountains were reputed to be fearsome, ugly environments inhabited by uncouth peoples engaged in such lowly activities as subsistence farming, mountain pasturing, and trade over the highest European passes. The mountains were strictly excluded from Grand Tour itineraries until the late eighteenth century; they did not coincide with the aesthetic canons of the time, conditioned by the perfection of classical forms. The Alps and European mountains generally were seen only as a barrier that had to be crossed from time to time.[43]

A change in perception of the mountains and mountain tourism came about in the late eighteenth century, characterized by three main aspects:

- a move away from the classical ideal, replacing the perfection of forms with the appeal of randomness and chance;
- the progress ushered in by the scientific revolution, with the affirmation of the natural sciences, recognizing the mountains as an environment to be discovered and studied for its flora and fauna, geology, and the economic value of its metals;
- Romanticism and Rousseau's myth of the good savage, linking to the ancestral, positive attributes of the mountains as a pristine realm uncorrupted by the habits of city life.

The interest in mountains in the eighteenth and early nineteenth centuries was primarily scientific, mainly focusing on natural ecosystems. But a spirit of adventure also began to grow, with the intrepid drawn to scale the highest peaks, also driven by the romantic allure of the unknown. In particular, tourists from Western and Central Europe established the mountains as a desirable destination, with the British as major promoters. An adventuresome spirit thrived in the relatively unfettered elite classes, whose passion for climbing originating from a love of risk. The first summit to be conquered was Mont Blanc in 1786, after twenty years of failed attempts. However, the most ardent climb, its epic laced with tragedy, was the Matterhorn in 1865 after thirty years of failure.[44]

A nearly indispensable reference for descriptions of natural mountain ecosystems is the texts of Swiss botanist and geologist Horace-Bénédict de Saussure (1740–1799), a very important figure in the birth of mountaineering and mountain tourism in the modern sense. It is therefore interesting that it was a naturalist, that is, one approaching the Alpine environment with research and cataloguing objectives first and foremost, who initiated the favorable perception of the mountains that we know today. Under his father's and his uncle's influence, the adolescent Saussure had developed a keen interest in the plant kingdom, which took him to the Alps on numerous occasions to study their flora.

In 1760, he determined to calculate the altitude of Mont Blanc, which was not then known, and promised a reward to the first person to find a route to the summit. On August 8, 1786, Jacques Balmat and Michel Gabriel Paccard pioneered the Grands Mulets route and were the first to set foot on the summit of Mont Blanc. A concierge doctor in Chamonix, Paccard was passionate about not only medicine but also botany and mineralogy. These passions soon brought him into the company of Saussure, and the idea took the form of climbing its summit precisely to further his studies. Balmat, however, was a local hunter and guide who used to collect crystals. Thanks to the route opened by Balmat and Paccard, Saussure—accompanied by seventeen guides plus his servant—reached the summit of Mont Blanc on August 3, 1787, where he immediately had a tent erected before proceeding to calculate the height.[45]

Saussure's description of seven Alpine journeys and his scientific observations are published in four volumes titled *Voyages dans les Alpes* (Travels in the Alps). An admiration for these natural ecosystems is evident, especially in his description of the Mont Blanc area, presented as a magnificent picture with high spires, massive glaciers, and imposing massifs,[46] as well as awe-inspiring sunsets among the Alpine peaks:

> But the beauty of the evening is the magnificence of the spectacle of the setting sun. The haze in the air, like a light gauze, tempered the brilliance of the sun and cloaked the immense expanse we had below our feet; a most beautiful crimson belt girded the western horizon while to the east the snows at the base of Mont Blanc, colored by this light, presented the greatest and most unusual spectacle.[47]

Among his other travels, Carlo Amoretti also showed a particular interest in the natural ecosystems of mountain areas. Examining the account of his travel experience to Lake Maggiore in northern Italy, we find several references to the alpine arc. But Amoretti's main emphasis is the interaction between human society and the environment, commenting on the upland cattle pastures in use from June to late September.[48] The English painter and writer William Brockedon (1787–1854), who visited the Alps numerous times in the 1820s, also acknowledged the liveliness at the Splügen Pass between Italy and Switzerland, with the Bergamasque shepherds and their 'immense flocks of sheep'.[49]

Amoretti, therefore, wrote careful observations on the interaction of farming and mountains, giving insight into how the natural Alpine ecosystem interacted with humans. Also capturing his attention were the mountains in the vicinity of Lake Como. He described their rocky peaks and the large caves where their waters pooled. Particularly awakening his interest was the presence of seashell marbles and fossil seashells of all kinds (especially ammonites), proving that the Po Valley was covered by a sea in prehistoric times, referring to the prevailing theoretical framework at that time.[50]

In contrast, the Scottish journalist and travel writer Leitch Ritchie's (1800–1865) interest in the Alps was focused on the challenges of human life in mountain ecosystems. He wrote an account of his travels in the alpine area titled *Travelling Sketches in the North of Italy, the Tyrol, and on the Rhine*, published in 1832 and accompanied by beautiful engravings by painter Clarkson Stanfield.[51] Ritchie's description of the village of Simplon, in Switzerland near the pass of the same name, is apt:

> In this place, the rays of the faint sun are intercepted, for many months of the year, by the mountains, and the neighboring glaciers breathe like death upon the spot: yet here stands a society of human dwellings, called the village of Simplon.... They bring their provisions from Italy or the Valais; for even the hardy potato is but little successful in its struggles with this dreary climate.[52]

In this case, the rugged beauty of the mountains represents a context of serious difficulty for human society, which, from Ritchie's account, appears unable to fully adapt to the Alpine ecosystem and its harsher aspects. Perhaps, however, Ritchie's perspective was influenced by the fact that he transited through the area and did not pause to analyze all aspects of it in minute detail, as we have seen with other travelers.[53]

Indeed, there are beautiful texts describing in detail the Alpine ecosystems observed by the writer on treks or climbs. Take the case of English clergyman Samuel William King (1821–1868), not only the rector of Saxlingham Nethergate, Norfolk, but also an enthusiastic entomologist and geologist who traveled frequently on the continent and was an avid mountain climber. He was usually accompanied by his wife, Emma Fort, who belonged to a wealthy family of calico printers in Manchester with some members in political office. Accounts of a long expedition made by the couple around 1855 are contained in King's only book, *The Italian Valleys of the Pennine Alps*, published in 1858.[54]

King describes settings that go from gentle and pleasant to majestic and wild. He describes Val Ferret, one of the valleys closest to the Mont Blanc massif, and the trail that winds pleasantly amid rhododendrons and venerable larches decorating the slopes as if in a park. In contrast, in the mountains near Courmayeur, accompanied by a guide, King speaks of great glaciers and fields of eternal snow looming over them:

> Niched in the broad rifts of the stern precipices, whence the ice-streams and débris are shot in innumerable furrows down the polished rock, these gigantic sheets of ribbed and crevassed ice, of dazzling whiteness, streaked with blue chasms, hang as if cut off at an immense elevation above the valley, and ready to fall into it.[55]

A peculiar aspect of King's experience and related work regards the potential of the resorts in Valle d'Aosta on the level of tourism proper. Indeed, in the context of the natural ecosystems by which he is surrounded, he provides an extraordinary and pioneering definition of Alpine tourism:

> Few spots in the Alps have left with us more vivid recollections than Courmayeur, surrounded as it is by the most magnificent mountain scenery in Europe, with commanding points of great elevation from which to scan it with an ease and familiarity rarely attainable. The geologist, the naturalist, the artist, or sportsman, or anyone in fact who has higher aims and pursuits than the mere superficial tourist, may find at Courmayeur and in its neighbourhood ample occupation and enjoyment for a long sojourn.[56]

Another mountain ecosystem that has attracted the attention of tourists since the eighteenth century is the Dolomites, in Trentino-Alto Adige, Veneto, Friuli, and Austria. The Dolomites are named after the French naturalist Déodat de Dolomieu (1750–1801), who first analyzed the predominant rock in the area. In 1864, the volume *The Dolomite Mountains*, a travel account by Josiah Gilbert and George Cheetham Churchill, was published.[57]

Gilbert (1814–1892) was a draftsman and watercolorist, and Churchill (1822–1906) was a former lawyer who had converted to the natural history. Together, they helped radically transform our knowledge of the Dolomites.[58] Armed with notepads and paints, they explored that still unknown and wild part of the Alps in the 1860s. They loved to venture out to discover those regions together with their wives. Indeed, the teamwork among the two couples opened quite a few doors for them during the field research for *The Dolomite Mountains* and facilitated relationships with the local inhabitants. In the introduction to the book, they described themselves as 'two holiday-making Englishmen, each accompanied by his wife', which clearly suggests that their interest in the mountains had now shifted from that of purely scientific and naturalistic travel to that of experiential travel, personal growth, and the search for new excitement (although Churchill was a fully respectable botanist and geologist). They took long journeys on foot, sometimes by horse or mule, and rarely by cart. They preferred paths to roads and haylofts to beds, relishing the fatigue and cold and unlikely foods eaten from wooden bowls as they set out to discover natural wonders, in which a couple of generations of erudite souls had learned to seek the Picturesque and the Sublime: lofty peaks, cliffs, glaciers, gorges, waterfalls. The two men appreciated the diplomatic acumen of their wives, who helped them interact with the locals, especially when it came to finding shelter in rural dwellings. In addition, the four travelers systematically divided the activities of documenting the natural ecosystem, assigning to each other on a rotating basis the tasks of writing, drawing, and reading documentary sources.[59]

In the introduction to their book, Gilbert and Churchill note how they were drawn to this area by reading the following passage from the Murray guidebook:

Here the traveller obtains a view of the Dolomite Mountains. They are unlike any other mountains, and are to be seen nowhere else among the Alps. They arrest the attention by the singularity and picturesqueness of their forms, by their sharp peaks or horns, sometimes rising up in pinnacles and obelisks, at others extending in serrated ridges, teethed like the jaw of an alligator; now fencing in the valley with an escarped precipice many thousand feet high, and often cleft with numerous fissures, all running vertically.[60]

During their eight-week stay in 1861, with over two hundred miles traveled, they noted that they had not encountered a single tourist, English or otherwise.[61] This highlights the fact that the Dolomites, now an important Italian and international tourist destination, were then a sparsely developed environment. In 2009, the Dolomites were declared a UNESCO World Natural Heritage Site for their aesthetic and landscape value and scientific importance at the geological and geomorphological levels. The UNESCO Dolomites Foundation was created shortly thereafter in order to safeguard and protect these unique places from human overuse. The unique landscape is dotted with picturesque and well-kept villages. The small houses have wooden decorations and balconies overflowing with flowers, and all have steep roofs to prevent an excessive weight of snow from accumulating. Two well-known pearls in these mountains are the resorts of Cortina d'Ampezzo and Corvara in Badia, both drawing in luxury tourism.[62]

Interest in the natural landscape of the Dolomites thus grew quickly over time. Tourist demand is very high today both for summer and winter vacationing and includes the pursuit of wellness. We read on the Official Website of the Dolomites UNESCO World Heritage:

The Dolomite landscape can be separated into its main landscape units in order to emphasise those elemental features that are most recurrent and recognisable throughout the region. The beauty of these landscape units is the product of the intimate relationship between its genetic and geological origins, its morphological structure and the nature of its topsoil.[63]

There are many other resorts in this region that welcome tourists from all over the world all year round. In the winter, dotted with quaint and cozy lodges, the Dolomites are a popular destination for the ski slopes and a venue for many international events. In the summer, they are a popular hiking destination among peaks and placid Alpine lakes, offering spectacular views.

Returning to Gilbert and Churchill's experience, before reaching the Dolomites, they had an early encounter with the mountain ecosystem during a hike on the Gamskarkogel in Austria. The ecosystem on this mountain was characterized by a wind that cut across the mountain ridge like a scythe. From there, however, there was a superb view of the mountains, which were sometimes obscured by powerful storms that filled their recesses with terrible darkness: 'Glaciers clung to their sides, "plastered" close under their peaked summits, and hanging down in tongues, from which in glistening streaks the torrents poured below'.[64]

Gilbert and Churchill then dwell at length on their description of the Dolomites and their geographical, geological, and natural features, providing a clear picture of this natural ecosystem. The Dolomites amaze the observers because 'they rise with such lofty independence of the surrounding scenery, are shattered into shapes so strange, cut the sky with such sharpness of outline, and gleam with so unearthly a light, that you are riveted by the spectacle'.[65] The authors dwelled on the description of mountain ecosystems in the area of the Marmolada, the highest peak in the Dolomites, describing many plants and flowers and observing how exposure caused vegetation to change even in the space of a few hundred meters and how flora adapted to climatic conditions or the presence of water.[66]

In general, Gilbert and Churchill's *The Dolomite Mountains* offers a great deal of information about tourism in Alpine ecosystems, allowing us to note, in effect, that the name of this area itself is a result of science-based ecosystem tourism. Thanks to the publication and wide circulation of this book, a hitherto unknown Alpine area well off the tourist's beaten track became known internationally by its current name.

We find other descriptions of the alpine environment in the writings of William Martin Conway (1856–1937), professor of art at Liverpool and Cambridge universities and English politician. He undertook a series of journeys in which climbing and a passion for the mountains were focal points. In 1894, he made an extensive mountaineering campaign from one end of the Alps to the other. In 1896, he traveled to the Svalbard Islands in the Arctic Ocean off the coast of Norway, explored their shores, and made numerous ascents. In 1897 and 1898, he turned to the Andes of Bolivia, making the ascents of Illimani and Sorata (without reaching the summit). He also attempted Aconcagua and Mount Sarmiento in Tierra del Fuego. Conway was the author of a number of books describing his travels in the mountains of Europe and the Americas. He describes the beauty of the natural ecosystems, dwelling on the awe they inspired in mountaineers. The turn of the twentieth century was a fertile period for him, with titles that exemplified the new allure of the mountains that Conway intended to convey to his audience through descriptions of his experiences and the scenery he had seen. Examples include *The Alps from End to End* (1895), *The Bolivian Andes* (1901), and *Aconcagua and Tierra Del Fuego: A Book of Climbing, Travel and Exploration* (1902).[67]

Conway provides accurate descriptions of the Alpine scenery in *The Alps from End to End*. While traversing the Italian Alps in 1894, he was as fascinated with Mont Blanc, as King before him, especially by the crevasse-furrowed glaciers illuminated by the dazzling sun and the contrasting walls and ravines shrouded in gloomy darkness.[68] At one point, Conway could not help but pause enraptured:

> I halted to gaze on the wondrous panorama, thus astonishingly revealed. Assuredly nowhere else is Mont Blanc better seen than from this Ruitor névé. No foreground more admirably serves to set off its blue shadowing buttresses and cream-coloured domes than the flat white area of this magnificent snow-field.[69]

2.4 Worlds of Ice

The tourist appeal of ice has a very long history, originating in scientific voyages and explorations near the polar circles in the eighteenth and nineteenth centuries. The Scandinavian Peninsula was an early focus of explorers of cold regions. Thousands of years of advancing and retreating glaciers have left marks that are still clearly visible in the landscape. Winters are extreme, with temperatures north of the Arctic Circle far below 0°C, creating an ecosystem dominated by ice.[70]

These natural frozen climes fascinated a young Italian traveler, Giuseppe Acerbi (1773–1846) at the end of the eighteenth century. He was born near the town of Mantua, northern Italy, into a wealthy family. He studied in Mantua and graduated in law from the University of Pavia. Shortly after graduation, in 1796, the twenty-three-year-old Acerbi began an educational journey that covered much of continental Europe and England. His next itinerary, in 1798–1799, took him to Denmark and Sweden, reaching the North Cape with an expedition led by Swedish general Anders Fredrik Skjöldebrand.[71]

Acerbi published a two-volume account of his journey, illustrated with engravings, in London in 1802: *Travels through Sweden, Finland and Lapland to the North Cape in the Years 1798 and 1799*. The book was later translated into both French and Italian. It includes a magnificent description of the frozen sea between the modern-day states of Sweden and Finland. On March 18, 1799, Acerbi, Skjöldebrand, the Italian Giuseppe Bellotti, and some guides set out from Stockholm, arrived at Grisslehamn on the Swedish coast, and, after crossing the Gulf of Bothnia still covered by ice, arrived at Åbo on the Finnish coast. The crossing was then possible by horse-drawn sleigh.[72]

About the organization of the crossing Acerbi writes: 'When a traveler is going to cross over the gulf on the ice to Finland, the peasants always oblige him to engage double the number of horses to what he had upon his arriving at Grisslehamn'.[73] Acerbi's group used eight sleighs, being three in company and two servants and having to face on the ice a route of forty-three English miles, thirty of which were without touching land. Acerbi describes the passage through this ecosystem as a most unique journey and the most astonishing sight a southern traveler could hope to see. He thought it would be a monotonous and dangerous journey, but instead, an incredible scene opened to his eyes. The ice sheet on the sea, initially smooth and even, became increasingly broken and irregular. As they proceeded, it took on an undulating appearance, evoking the waves of the sea. Eventually, they encountered ice boulders piled on top of each other, and some of them looked as if they were suspended in the air. To Acerbi's eyes, they appeared as an immense chaos of frozen ruins, adorned with superb ice stalagmites of a blue-green color.

Throughout their journey, Acerbi and his companions encounter no other living creatures, whether humans or birds. Acerbi writes: 'Those vast solitudes present a desert abandoned as it was by nature'.[74] Only the wind creates a rustling sound that is often accompanied by loud crunches from the chunks of ice that it dislodges and carries over great distances. The only

Figure 2.2 Winter crossing the Gulf of Bothnia on ice. Giuseppe Acerbi, *Vues de la Suede, de la Finlandie, et de la Lapponie, depuis le détroit du Sund jusqu'au Cap Nord* (Paris, 1803), plate 4.

animals that Acerbi believed inhabited this ice ecosystem were sea calves that could brave the rigors of the bitterest season. All they needed was an open hole in the ice near them for rapid escape into the water. The first island the group reached was Signilska, one of the small islands scattered in this part of the gulf, an archipelago named Aland.

Acerbi provides us with two different levels of analysis of the Finnish natural ecosystem. On one hand, there is a focus on the ecosystem understood as a set of classifiable elements of nature. Acerbi is an objective observer of the world around him and tends to categorize it. This classificatory approach is nourished by the scientific information that abounds in his *Travels*, revealing a strong naturalistic streak. On the other hand, he calls attention to the landscapes of these ecosystems, where the tourist *ante litteram* is emotionally engaged and expresses his feelings about the environment around him.[75]

After returning from Scandinavia, Acerbi's career in the nineteenth century was divided between diplomatic assignments, cultural management engagements, and directing his own agricultural estates. In particular, from 1825 to 1834 he was consul general in Egypt of the Habsburg Kingdom of Lombardy-Venetia and, from his office in Alexandria, devoted himself to a careful analysis of that country's agriculture, industry, and trade. These were the years of Egypt's economic flowering under Mehmet Ali and of geologist Giovanni Battista Brocchi's travels there (see Chapters 1 and 3). When Brocchi died of dysentery in Sudan, it was Acerbi who announced the news to Italy in a letter later published in various newspapers.[76]

American Charles Francis Hall (1821–1871) undertook a number of Arctic expeditions in the second half of the nineteenth century. As a young man, he lived in Ohio, where he held various jobs including blacksmith, journalist, stationer, and engraver, before developing an interest in travel. His first expedition to the Canadian Arctic zone was in 1860–1862. The main objective was to find evidence in the Inuit oral tradition about the explorations of English navigator Martin Frobisher in the 1570s. Thus, communication with locals was a central aspect of Hall's journey. He also intended to gather more information about the tragic fate of the 1845 British expedition to find the Northwest Passage, led by John Franklin.[77]

The two ships in the Franklin expedition, HMS *Erebus* and HMS *Terror*, had disappeared with their crews in 1847. British and U.S. expeditions had been undertaken in subsequent years to locate any survivors or at least to reconstruct what had happened. The tragic end of Franklin and his crew that had been pieced together through Inuit accounts gathered by British navigators was highly controversial in Victorian society. It involved the two ships being stuck in the ice, the crew being driven to the point of cannibalism, and ultimately perishing in the ice. The wrecks of the two ships were not found until 2014 and 2016.[78]

Hall began to collect all the material he could find on the Arctic landscape and survival in it. At the same time, he sought financial support for his own expedition project. In 1860, he disembarked from the whaler George Henry (which had given him passage from the United States) at the southern end of Canada's Baffin Island and spent two years exploring the area. Based on this early experience he wrote *Arctic Researches and Life among the Esquimaux* (1865).

Another expedition began in 1864 from the northern end of Hudson Bay, marking the beginning of five years and 3,000 miles of sleigh travel. Hall's last venture was to command a U.S. government-sponsored expedition to reach the North Pole. On June 29, 1871, he set sail from New York City aboard the steamship *Polaris*. He passed through the Kennedy and Robeson channels, which separate northwestern Greenland from the northeastern Canadian Arctic; charted both coasts; and reached the northernmost limit of exploration by ship. Hall died suddenly and under unclear circumstances on the return voyage.[79]

While ostensibly setting out to analyze and describe natural ice ecosystems, Hall's passion for these lands and his explorations is evident from the very first pages of his *Arctic Researches*:

> Approaching the north axis of the earth! Ay, nearing the goal of my fondest wishes. Every thing relating to the arctic zone is deeply interesting to me. I love the snows, the ices, icebergs, the fauna, and the flora of the North! I love the circling sun, the long day, the arctic night, when the soul can commune with God in silent and reverential awe! I am on a mission of love.[80]

Hall's excitement was great at the sight of the first iceberg, having longed for it for so many years. Having spotted it from a great distance, it was already nightfall when the expedition drew close. In that light, the iceberg looked like an alabaster mountain resting quietly on the bosom of the dark blue sea. On another occasion, Hall decided to enter into direct contact with an iceberg. Accompanied by an officer in the tender, he approached a large one floating alone not far from the ship. With some difficulty, he managed to climb onto it and then up to the top. He observed a topography characterized by pinnacles, ravines, gorges, and deep hollows. Above all, he noted that the iceberg is undergoing a process of decay and would evidently soon break apart. He wrote that the 'bergs have been known to approach and recede from each other in as beautiful and stately a manner as partners in the old-fashioned, courtly dances of years gone by' and that they are unpredictable in movement and deceptive in size.[81]

On November 19, 1860, the ice had spread across much of the bay located on the southern coast of Baffin Island, where the expedition ship was located, completely enclosing it and isolating it from the rest of the world.

They could not approach land, nor could anyone from the shore approach them until the ice became solid enough to allow them to cross. On January 10, 1861, having a great desire for exploration and, in particular, to observe real life among the Inuit, Hall decided to venture on a dogsled trip near Cornelius Grinnell Bay.[82] We discuss the details relating to the organization of this sled trip and the difficulties in Chapter 3. In particular, Hall's accounts focused on the interaction between the Inuit and the natural ecosystem around them and their ability to adapt to all eventualities.

Hall also witnessed the spectacle of the northern lights. In December 1860, he sketched the phenomenon in detail. In March 1861, he reported in his journal: 'It seemth to me as if the very doors of heaven had been open to-night, so mighty, and beauteous, and marvelous were the waves of golden

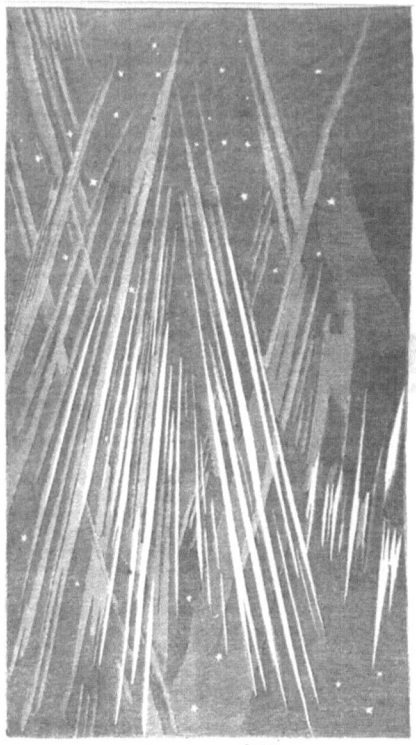

AURORA, DECEMBER 17, 1860.

Figure 2.3 Northern lights observed by Charles Francis Hall on December 17, 1860. Charles Francis Hall, *Arctic Researches and Life among the Esquimaux: Being the Narrative of an Expedition in Search of Sir John Franklin, in the Years 1860, 1861, and 1862* (New York, 1865), p. 150.

light that a few moments ago swept across the "azure deep", breaking forth anon into a flood of wondrous glory'.[83]

Ice worlds were also present at more accessible latitudes, obviously in a less extreme context than that described by Hall in lands near the Arctic Circle. The aforementioned Samuel William King made frequent reference to ice in his 1858 volume *The Italian Valley of the Pennine Alps*. Such was the case with his visit to Valtournenche, a side valley of Valle d'Aosta, where he was greeted by a majestic panorama of mountains and glaciers. Light figures very prominently in his description of the massive glaciers that loomed over the valley.

> The glowing rays of the rising sun soon shot up behind the snow-crest, and quickly dispelled the cold grey tints of early morning, fringing the ridge with a singularly defined edge of brilliant light—until the great orb suddenly uprose, red and frosty-looking as on a January morning.[84]

As King continued with the guides toward Theodul Glacier, located on the Swiss side of the Pennine Alps (although its upper basin touches the Italian region of Valle d'Aosta), he was fascinated by the ecosystem in which he found himself. With the guides, he climbed a huge old moraine that indicated to him a much larger extent of the glacier in earlier times; the environment around it appeared to consist of a mixture of coarse gneiss, limestone, and shale, with some serpentine. The terrain was hard-packed with ice and promised a favorable ascent: at this point, they dismounted from the mules to continue on foot as caution was absolutely imperative on this hike. The above-freezing temperatures in this season caused the surface of the glacier to change every day and new crevasses to open. The crevasses were a deep abyss of blue ice walls, fringed at the top by a row of beautiful icicles many meters long. In the late morning, everything around them seemed to change, for the sun with its warmth began to thaw the surfaces and made it difficult for hikers, who sank ever deeper into the snow.[85]

King reached the upper part of the glacier, which was shaded from the sun by the tops of the Breithorn and Klein Matterhorn. The high altitude, however, produced dizziness, fatigue, difficulty breathing, and violent palpitations in many of the travelers (even in the guides). In any case, from that height, almost 11,000 feet, the view was superb and some magnificent ranges could be seen above the nearby valleys. But it was the Matterhorn that caught King's attention, with this peculiar form of absolutely inaccessible solid rock obelisk. The icy ecosystem, however, was not devoid of life, and King notes the presence of solitary specimens of *Ranunculus glacialis* L. in bloom. In addition, the only representatives of animal life were snow fleas as black and shiny as grains of gunpowder and a few aphids.[86]

Another glacier visited by King was the Lys Glacier located on the south side of Monte Rosa in Italian territory. Through his observation of this glacier, King made some important points about the great fragility and transformative capacity of these ecosystems. He wrote that the Lys Glacier had recently made extraordinary advances and was thus plowing its way downward until it approached the ancient frontal moraines in the valley. The glacier, which had retreated in the first half of the nineteenth century, was instead advancing to the point where, in 1854, it made the astonishing advance of 200 feet downward with an average thickness of 50 feet. King noted: 'The increase of one glacier and decrease of another, even on the same range of mountains, is most puzzling'.[87]

The description of the glacier itself seems worthy of a work of art:

> The glacier wall which stretched to the right of the *Nase* above us was a magnificent object, and, along its crest, huge peaked icebergs, light spiry pinnacles, and massive blocks, of a pale transparent green ice, stood out boldly against the sky, at a height of from 2,000 to 3,000 feet above us. The snow was brilliantly pure, and, with the glare of the sun, almost blinding. This amphitheatre of precipitous glaciers and intervening piles of rock closed up the head of the vast Felik Plateau, which we were traversing along the course of a medial moraine, which had its origin at the base of the *Nase* and was distinctly marked in its sinuous descent for miles down the glacier.[88]

Ice worlds—whether they were the frozen seas of Scandinavia, the inhospitable lands of extreme North America, or the more accessible but equally dangerous glaciers in the Italian Alps—aroused the great fascination of travelers in the eighteenth and nineteenth centuries. At times this was interwoven with an interest in mountain ecosystems, as in the case of King in this chapter or Mary Wollstonecraft Shelley in Chapter 1. The complexity of polar and ice-intensive mountain ecosystems is evident in the accounts we have analyzed, characterized by a delicate balance between local populations, a particularly harsh environment, and scarce food resources, as we read in Hall's accounts. This fragility was also noted somewhat closer to home, in King's exploration of Alpine valleys, whose ice and snow provide a habitat for snow fleas, still an important component of these ecosystems.

2.5 The Allure of Deserts

Natural desert ecosystems are areas characterized by an arid environment and scarce rains, making them one of the driest biomes on our planet and difficult to inhabit. These climatic characteristics, added to the infertility of its soil, which is mainly composed of desert sand or rocks, result in barren, lonely landscapes where few animal or plant species are able to dwell. There are hot deserts, such as the Sahara, and cold ones, such as Antarctica and high mountain peaks. Deserts comprise a variety of ecosystems, there being many different types of deserts: deserts in regions of trade winds or tropical winds, mid-latitude deserts, deserts caused by permanent barriers to moist air flows, monsoon deserts, and coastal deserts.[89]

Although ostensibly hostile environments, deserts have always been traversed for reasons of trade and have been tourist destinations since ancient times. This type of environment has always attracted the imagination and stimulated the desire for discovery in many travelers. In the eighteenth and nineteenth centuries, numerous European travelers crossed the Middle East and published travel diaries, accounts, and reports. A great deal of information can be obtained from these texts, not only about archaeological sites, often the principal reason for the journey, but also of a geographical and ethnographic nature and, to a lesser degree, about the natural ecosystems of these areas. In this section, we focus on hot deserts.

The first desert ecosystem record we analyze is that of Frenchman Louis Damoiseau (1775–1832). He was born into a modest family in Chartres and studied veterinary medicine at one of the most important European schools, that of Alfort. He later practiced his profession in his hometown, then at the French national stud farm in Le Pin-au-Haras. In 1818, on assignment from the Ministry of the Interior, he participated in an expedition to the Middle East with the aim of acquiring Arabian horses. He remained in Asia Minor until 1821, visiting those lands and later, back in France, publishing his reports as *Voyage en Syrie et dans le désert*.[90]

Damoiseau had the chance to describe the desert ecosystem of Syria. To see the horses of a desert tribe he set out from Aleppo in mid-June 1819 with a caravan. Local companions rode on camels or mares, and some owned donkeys, which they loaded with barley. Damoiseau started on a camel, which he said was in rather bad shape, but was fortunately soon able to obtain a horse. The caravan marched through a desert ecosystem and had to travel long distances between one water source and another. Damoiseau made frequently comments on the discomfort of travel in such climes and emphasized the oppressiveness of the heat especially for European travelers, who were not used to such temperatures and the desert wind that blew continuously and made the journey even more uncomfortable. In addition, Damoiseau and fellow travelers were tormented by thirst, which was exacerbated by the mirages that were always appearing before their eyes and making them long for a nearby lake where they could quench their thirst.[91]

Nonetheless, there were some green oases in the desert, bringing great relief to Damoiseau after the most difficult stretches. This is how he described an oasis probably at the foot of Tell es-Sultan, a hill near modern-day Jericho:

> What joy we felt when, as we approached, we saw a beautiful meadow covered with a thick and bushy grass, and whose beautiful green

contrasted in the most pleasant way with the bare and burning ground that we had just crossed![92]

Damoiseau was especially delighted by a clear stream of water that ran through the green and was screened by a lush reed thicket, in which many wild boars hid. This was not an unspoiled corner, however, given the frequent caravans Damoiseau witnessed relying on that 'islet of greenery' and the spring at the foot of the hill. Not coincidentally, Tell es-Sultan has been identified with Ancient Jericho and is an important archaeological site that is a candidate for UNESCO heritage status. Due to its 'diverse geological formation and unique tropical and sub-tropical climate zones, alluvial soils, and perennial springs', the area has attracted hunters and later farmers since earliest antiquity as a place to settle.[93]

Along the way, often in the vicinity of the wells, Damoiseau and companions also encountered other caravans, such as a group of Arabs belonging to a tribe that used to camp in the part of the desert where the ruins of Palmyra are located. This tribe lived in the desert ecosystem and traded soda and rock salt with Aleppo. Damoiseau was fascinated by these environments, so far removed from those in which he was accustomed to living. He was also extremely interested in the horses he saw, which he readily called his real objects of desire. He also noted how life seemed to be present only in the areas where there were wells, whereas the rest of the desert seemed entirely bereft of it, except for caravans moving for commercial, military, or religious reasons.[94]

Regarding desert ecosystems, also exemplary are the pages of the English explorer William George Browne (1768–1813), who traveled to Egypt and the Middle East on several occasions beginning in the 1790s. For example, in 1793 he visited Sinai and, joining the great caravan that traveled the desert route each year, reached Darfur. On a later journey in 1812 and 1813, his goal was to reach Samarkand and explore the regions of Central Asia. However, shortly after leaving the city of Tabriz, in present-day northwestern Iran, to reach Tehran, he was murdered.[95]

One of the publications Browne wrote about his journeying, *Travels in Africa, Egypt, and Syria from the Year 1792 to 1798*, contains a description of the Egyptian desert. On his first trip, he had a miserable experience in the desert near the oasis of Siwa, which he reached on horseback with an interpreter and guides. The heat, sand, and undrinkable water were characteristic features of this environment and had strongly negative impacts on travelers. Indeed, Browne was stricken with dysentery and fever.[96]

He also provides us a vivid description of the desert of Upper Egypt on his journey to Darfur (in today's Sudan). A rise afforded him a commanding view of an immense valley, almost entirely composed of rocks and sand, even though 'diversified by small bushes of the date tree, and other marks of vegetation, near the spring where we designed to repose'. It was the most barren land he had ever seen, although such hostile ecosystems were regularly traversed by merchant caravans. Keep in mind that Browne and his caravan followed a route through oases, towns, and villages. He mentioned a 'Great Oasis', perhaps a reference to the major Kharga Oasis. The economies of the settlements encountered by Browne and companions along the way were based on water wells, contact with caravans, the sale of products such as dates, and, as in the case of Bir el-Sheb, the extraction and trade of alum.[97]

Two travelers we introduced in Chapter 1 also commented in their writings about the desert environments they encountered in their travels: the French philosopher and orientalist Count de Volney and the Italian geologist Giovanni Battista Brocchi. They complete this brief analysis of traveler experiences in hot deserts that we began with Damoiseau's and Browne's accounts.

Volney was a member of a social class that had sufficient wherewithal to travel for personal reasons far before the advent of mass tourism. What drew him to Egypt and the Middle East was a taste and passion for learning and the discovery of the sites of some of the most important civilizations of antiquity. He repeatedly analyzed desert ecosystems in his 1783–1785 travel report, published in 1787 under the title *Voyage en Syrie et en Égypte*.[98] His accounts emphasized the presence of hot, dry winds in the Egyptian desert. He also noted general aspects of the ecosystem—including the climate, the geomorphology of the land, and the presence of plants and animals—that conveyed the complexity of these lands and the difficulty for humans to function in them.[99]

Giovanni Battista Brocchi's *Giornale*, published posthumously in several volumes in the 1840s, also repeatedly notes how the desert ecosystem in Egypt begins where the influence of the Nile diminishes in the lands adjacent to it. Brocchi described an expanse of sand dotted with outcrops of siliceous sandstone and noted how the vegetation was limited to a few shrubs and a few red acacias. As a geologist and naturalist, he had a tormented fascination for these environments; although overwhelmed by the barrenness of the place, he reported valuable information about the ecosystem, recording his observations of the very sparse vegetation and fauna in addition to geological aspects.[100]

Desert ecosystems are composed of a variety of elements that are not immediately apparent to the eye. They change with the seasons, between day and night, by their distance from rivers, and in many other ways. Since deserts are challenging habitats, the human experience is often associated with travel across them, with caravans and the animals that have made them possible for centuries. Another aspect we note in these accounts is that while deserts are some of the most inhospitable places on our planet, they are also some of the most mysterious and suggestive. They have found a place in our collective imagery through legendary tales spanning the centuries. Oftentimes the deserts we know now were green and fertile at some point in the past and hide vestiges of ancient civilizations waiting to be discovered. In general, the case studies we have examined in this chapter and in Chapter 1 evidence the desire of many travelers—some perhaps more reluctant, such as Volney—to become acquainted with these environments, often putting themselves to the test in extreme situations.

2.6 Plying the Oceans

The marine ecosystem, like any other, is a self-regulating mechanism composed of organisms and their environment in a symbiotic relationship. The richness of a marine ecosystem is a function of the diversity of living organisms (biodiversity), which is a measure of its health. The oceans are home to many forms of animal and plant life, which are grouped into four biotic macro-groups that represent all life in marine aquatic environments. These are neuston, organisms that are on or directly below the water surface; plankton, organisms and microorganisms that are unable to resist currents and are transported by water; nekton, motile organisms such as fish, cetaceans, turtles, and cephalopods; and benthos, organisms that live in contact with the floor of an ocean, lake, river, or other body of water.[101]

Oceans have long been an element in long-distance human travel, especially to reach other continents. For the most part, they are not the purpose of travel but an ecosystem to be traversed to reach a destination. Nevertheless, the travel accounts we have analyzed provide important information on marine environments. Interesting in this regard are the travels of Georg Heinrich von Langsdorff (1774–1852), who describes some of the marine environments he transited.

Langsdorff was born in Rhine Hesse. He studied medicine in Göttingen and developed a great passion for the natural sciences. As a surgeon and military doctor, he initially resided in Portugal and then participated in the first circumnavigation of the globe organized by Russia (1803–1806), traveling from Copenhagen to Kamchatka (1803–1805). He became a subject of the Russian Empire and was appointed consul general in Brazil in 1812. After relocating to Rio de Janeiro, he explored the surrounding region at length, studying its nature.[102]

Here we are interested in Langsdorff's descriptions of the sea during his various ocean voyages. His writings—penned in German and later translated into other languages—provide interesting information about the ocean flora and fauna. He noted differences between one area of the planet and another and recorded the feelings that these sights aroused in him. Setting sail from Copenhagen in September 1803 with the Russian expedition and bound for the Canary Islands, Langsdorff saw large numbers of dolphins, skipjack tunas, flying fish, sharks, whales, and 'numerous sorts of aquatic birds unknown to us before', as well as vast expanses of kelp floating on the waters. Impressed by this variety, Langsdorff declared: 'I am almost inclined to think a sea voyage not at all less amusing than a journey by land'.[103]

Reaching the ocean between the two tropics, Langsdorff found himself sailing through very hot days, in which the heat was bearable only with very light clothing, and cool nights, during which the captain obliged the men to cover themselves well while they slept on deck, careful not to underestimate the temperature change and the consequences for their health.[104]

After stopping in Brazil, the expedition took to the sea again in February the following year, headed for Cape Horn. Langsdorff and his shipmates found other details of the marine ecosystem that attracted their attention. They no longer saw the fish and dolphins that had characterized the ecosystem near the Canary Islands or tropical birds, but there were great numbers of albatrosses and petrels, patches of brown algae, and whales (at one point, the ship collided with a whale, the violent impact causing general fright among those on board).[105] Here and there, Langsdorff observed shimmering stripes and spots on the surface of the relatively still water that could be distinguished at a great distance. Langsdorff concluded that they were oily, fatty substances released by the whales as they breathed or defecated. The size of these patches, coupled with assumptions about their source, suggested that small amounts of oil were able to spread over a very large area. He concluded: 'The idea, which I believe originated with Dr. Franklin, that the waves of the sea, when violently agitated, might be stilled with oil, was probably borrowed from this circumstance'.[106]

Once again, the naturalist's scientific observations intermixed with the traveler's emotions in a new world of wonder. Langsdorff's accounts also described the hardships he experienced on the windswept sea. Skirting the coast of North America from Sitka, Alaska (then under Russian rule) to San Francisco in 1806, the ship of the Russian-American Company—a chartered company founded by Paul I, Emperor of Russia, in 1799 for the exploitation of Russian America—dropped anchor in Grays Harbor, a bay at the mouth of the Chehalis River in modern-day Washington. The captain then sent a baidarka with three men to explore the mainland, and Langsdorff joined the expedition.[107] Returning to the mother ship in the evening, the kayak encountered dangerous conditions:

> Scarcely had we passed the farthest surf when it became entirely dark, and the moon, on which I had relied for lighting us on our way, was completely overshadowed with clouds. At the same time, a strong south-south-west wind rose, and the ship, partly from the darkness of the night, partly from the high waves and the low manner in which our boat was

built, was not to be seen. We therefore rowed with all our might towards the place where we supposed she must be.... At length, about half past ten, notwithstanding the darkness of the night, the force of the wind and the waves, and being obliged to take a circuit to avoid the surf, we arrived safe and well at the ship....[108]

The sea in this case posed a danger, even though it was not an open ocean but rather a portion of the coast enclosed in a large bay. Nevertheless, the tension of that experience was described by Langsdorff with great intensity, and at the same time, we sense his fear of this faraway natural environment. Capping off the happy ending of the baidarka adventure, Langsdorff wrote how the crew aboard the mother ship welcomed him and his companions with joy: 'they had given us up as having become a prey either to the fury of the waves, or the savages upon the shore'.[109] The wind-whipped sea and the indigenous people of the coast—inevitably labeled 'savages' by the European traveler—were thus the embodiment of the fury of nature.

Other interesting accounts of the ocean as a backdrop for adventurous travel in nature are given to us by the Briton Maria Graham (1785–1842), giving us insight into the perception of marine environments on long journeys from Great Britain to India and beyond. Graham was the daughter of a former naval commander who was later appointed head of the East India Company's shipyard in Bombay. She met Captain Thomas Graham on the very voyage to India and married him as soon as they reached their destination. After his death in South America, she was later (and better) known as Lady Callcott for her second marriage to the painter Sir Augustus Wall Callcott. Maria traveled to India and Latin America, worked in 1824 as governess of the young Princess Maria, daughter of the Emperor of Brazil, and later visited Central Europe and Italy.[110]

In 1821, she crossed the Atlantic to Latin America on the HMS *Doris*, leaving England on July 31 and stopping in Madeira and Tenerife. She accompanied her husband Thomas, who was in command of the Doris and headed for Chile, where he was to protect British mercantile affairs in that area. Leaving Tenerife on September 1, Maria Graham recorded the following impression of the marine environment as the ship plied the waves of the Atlantic:

The flying-fish are become very numerous, and whole fleets of medusæ have passed us; some we have picked up, besides a very beautiful purple sea-snail. This fish has four horns, like a snail, the shell is very beautifully tinted with purple, and there is a spongy substance attached to the fish which I thought assisted it to swim: it is larger in bulk than the whole fish. One of them gave out fully a quarter of an ounce of purple fluid from the lower part of the fish. A fine yellow locust and a swallow flew on board; and as we believe ourselves to be four hundred miles from the nearest land, Cape Blanco, we cannot enough admire the structure of the wings that have borne them so far.[111]

The following year, in late March, we have a completely different situation. Circumnavigating the Southern Cone from Rio de Janeiro to Valparaíso in Chile on the *Doris*—not long before her husband died of fever—Maria Graham found herself in the middle of a storm. In this case, the description of the seascape is dark and foreboding, but it seems less tragic and more filled with admiration than Langsdorff's experience at Grays Harbor. In spite of the dark features, Graham is indeed drawn to and almost amused by the roaring adventure in the ocean:

> We are in the midst of a dark boisterous sea; over us a dense, grey, cold sky. The albatross, stormy petrel, and pintado are our companions; yet there is a pleasure in stemming the apparently irresistible waves, and in wrestling thus with the elements.... If I look abroad, I see the grandest and most sublime object in nature,—the ocean raging in its might, and man, in all his honour, and dignity, and powers of mind and body, wrestling with and commanding it....[112]

This epic description is contrasted with the ridiculous one Graham gives of the world inside the ship, particularly the great disorder caused in her cabin by the rolling waves. Not much separates the almost comically jumbled human realm from the sublime seascape. She also contrasts the 'frozen regions' to which they were heading after passing the Falkland Islands with the English environment they had left long ago, where people were waiting for spring and 'its gay light days and early flowers'.[113] In Graham's accounts, the ocean is an alternative to the human world and the security of everyday life in England. The beauty of the marine fauna, and the allure of a terrible storm, generates in her traveler's soul a love of the wild and especially of the varied marine environment.

Langsdorff's and Graham's accounts show how even the intervals between landfalls constituted a complete experience of immersion in a natural ecosystem, in this case the oceans with their diverse fauna and impressive spectacles, such as storms. We can therefore conclude by saying that transit by sea, as monotonous as it might seem compared to travel through more variegated terrestrial settings, in truth provided a very important experience of nature to the traveler, bringing a deeper understanding of a new environment.

2.7 Colonial Experiences

Colonial expansion had a significant transformative effect on the local environment. Between the fifteenth and nineteenth centuries, colonialism and industrialization enabled Europe to revive economically, culturally, and politically. However, although colonial and industrial activities did bring benefits, equally remarkable were the consequences for the environment. Ecosystems were radically altered both in Europe (in the case of industrialization) and in lands outside Europe (as a result of colonization). Colonialism exploited natural resources and local populations at different times, contributing to the transformation of the landscape and the loss of biodiversity. While colonialism produced benefits, mainly economic, for the motherlands, it also generated inevitable harmful consequences for the environment. The arrival of Europeans initiated a veritable environmental revolution, to the point where historian Alfred Crosby has spoken of ecological imperialism and the biological expansion of Europe.[114]

The introduction of new models of agriculture brought radical transformation to the colonial landscape, whether tropical, subtropical, or temperate. Traditional agricultural systems, based on manual and multi-crop farming techniques, were replaced with specialized and predominantly monocultural forms of agriculture. This created a system of food production for export to Europe. A number of recent studies, such as those by Helen Tilley, have focused on Africa. Tropical Africa was one of the last regions of the world to experience European colonialism. As a 'Living Laboratory', Africa is a far-reaching case study of the thorny relationship between the imperialist drive and the pursuit of knowledge (scientific, environmental, medical, anthropological) in the colonization of that part of the continent, where British efforts to transform and modernize were often unexpectedly subverted by economic and scientific considerations.[115]

As a general rule, traditional crops in the colonies were supplanted by plantations of products such as tobacco, tea, sugarcane, cocoa, rice, and cotton. Over time, however, the specialized crops damaged the soil, also causing the loss of biodiversity. The introduction into the colonies of European plant and animal species also disrupted local ecosystems. Forests, woodlands, mineral-rich soils, and animal-populated lands were intensively exploited to suit the commercial needs of the motherlands, which enriched themselves at the expense of the colonies. The exploitation of minerals and precious metals, which represented coveted resources for trade, was another cause of soil depletion. Agriculture also became linked to mining exploitation, as the substantial increase in cropland production stimulated the importation of cheap fertilizers.[116]

The centuries analyzed in this volume are characterized by colonial empires so we felt it was appropriate to include these diverse types of environments as well, which were undergoing profound transformations through European intervention. At the same time, as we mentioned in Chapter 1, naturalist expeditions were undertaken frequently in order to map the as-yet-unknown natural resources found in these parts of the world.

For example, French naturalist and officer Jean Baptiste Bory de Saint-Vincent (1778–1846) made an in-depth analysis of island ecosystems in the Indian Ocean between 1800 and 1802. In 1799 he learned of the imminent departure of a scientific expedition to Australia organized by the French government and obtained the position of chief botanist aboard one of the three participating corvettes, *Le Naturaliste*. He joined the expedition but left it after rounding the Cape of Good Hope because of conflicts with Captain Nicolas Baudin. It was then that Bory de Saint-Vincent set out to explore the islands of the Indian Ocean and, on his way back to France, the Atlantic island of St. Helena. Back in France, he published an *Essai sur les Îles Fortunées* that earned him election as a correspondent of the National Museum of Natural History and the National Institute, two of the most important scientific institutes in France, thus making him an active and prominent figure in Napoleon's scientific and cultural reforms. He was exiled during the Restoration but later welcomed back to Paris. He was also given roles of responsibility in new scientific expeditions, for example, to the Peloponnese, Greece, and Algeria.[117]

Having established the scientific authority of this naturalist, let us consider Bory de Saint-Vincent's descriptions of the environments he visited in the Indian Ocean. In Mauritius, with a young indigenous man, he climbed the monolith that towers 556 meters above the Le Morne Brabant peninsula at the southwestern tip of the island (and a UNESCO World Heritage Site since 2008 for the historical value associated with its having been an important refuge for Maroons in the eighteenth and early nineteenth centuries). The Le Morne Brabant peninsula is still a habitat for many endemic botanical species and the only place where the *boucle d'oreille* (*Trochetia boutoniana* F. Friedmann), Mauritius's iconic flower, can be found.[118] From the monolith of Le Morne Brabant, Bory de Saint-Vincent could enjoy a superb view of the lands near the coast, some of which were dotted with houses, cultivated fields, and beautiful gardens, while others were covered with pristine forests where nature maintained 'her primeval rusticity', an etymologically curious expression that well conveys the coexistence of natural and man-made landscape.[119]

Bory de Saint-Vincent observed an interesting forest ecosystem on Reunion Island, particularly along the Galets River, with thick vegetation that often posed an obstacle to his group. However, the naturalist took the long view of this redoubtable path, noting that

[t]he road we had chosen became very much entangled by ferns; the beauty of the greatest number of them, however, made me overlook the inconvenience they occasioned. As the places through which our route lay were extremely damp, and much shaded with wood, I discovered a great many cryptogamous plants, among which was a *Jungermannia viticulosa*

L., in flower, and a *Lycopodium* which very much resembled our *Lycopodium clavatum*.[120]

Equally interesting are the perspectives of the variety of environments in the American colonies provided by a number of travelers. For example, in Chapter 1, we analyzed prominent naturalist Alexander von Humboldt's account of the ecosystems of colonial areas during his travels through the Americas from 1799 to 1804. For Humboldt, the pristine nature of the tropics was something extraordinary, shaping the humans who inhabited it and of a beauty inconceivable to those who had not seen it for themselves. Despite this, in his travel accounts, Humboldt also made numerous references to the pleasantness of the rural scenery, where the effect of humans on the natural ecosystem was more pronounced, the latter inevitably tamed by the former.[121]

Another keen observer of landscapes and environments in the Americas is Maria Graham in her accounts of Brazil. She observed the beautiful Brazilian springs during her sojourns there in the early 1820s. She describes the scenery on a hike in modern-day Pernambuco State in October 1821:

Gay plants, with birds still gayer hovering over them, sweet smelling flowers, and ripe oranges and citrons, formed a beautiful fore-ground to the very fine forest-trees that cover the plains, and clothe the sides of the low hills in the neighbourhood of Pernambuco. Here and there a little space is cleared for the growth of mandioc, which at this season is perfectly green: the wooden huts of the cultivators are generally on the road-side, and, for the most part, each has its little grove of mango and orange-trees.[122]

Thus, the intermingling of the natural and human-made environment that Humboldt had already noted in his 1799–1804 trip is also found some twenty years later in the description of the resident Graham, demonstrating even in these small details the great changes brought by colonialism over the centuries to continents such as the Americas and Africa.

Shortly thereafter, in what is now the State of Bahia, Graham observed wild forests with all the splendors of Brazilian animal and plant life on display. She was enchanted by the flamboyant plumage of the birds, the brilliant hues of the insects, and the size, shape, color, and fragrance of the flowers and shrubs, most of which she was seeing for the first time. Again, however, nature and humans blended together in the newly dictated organization of colonial agriculture. With her companions, Graham visited a delightful 'pepper garden', enlivened that season by coffee blossoms. It was still too early for the equally beautiful cotton and pepper blossoms. One sensed the economic importance of these crops from the historical references Graham reported to readers of her *Journal of a Voyage to Brazil, and Residence There*. According to the information she gathered, not many years earlier Francisco da Cunha e Meneses had sent pepper from Goa to Brazil to be studied and acclimatized in Bahia. Later, when he was governor and captain-general of Bahia itself from 1802 to 1805, he expanded the experimental pepper crops, which were also extended to Pernambuco.[123]

The political and economic background of the importation of pepper to Brazil was part of the strategies of botanical exchange and acclimatization of allochthonous species from one colony to another within a given empire precisely during the eighteenth and nineteenth centuries. For example, Joseph Banks, president of the Royal Society of London, was requested by the British government to advise on cotton species to be cultivated in the West Indies. Banks also sponsored the study of breadfruit and attempts to acclimatize it to the British West Indies as a cheap source of food for the slaves.[124] On its part, France contributed to the acclimatization of 'useful plants' in the West Indies by transporting them by ship from the Indian Ocean colonies. The Botanical Garden and the Agricultural Society of Paris collaborated with French colonial structures to organize the smoothest crossings for the saplings and the seeds so that they would not suffer impacts, humidity, salt, sun, and the like on the ships.[125] The Paris Botanical Garden was also an important point of reference for Italian scholars in agricultural science, in particular starting with the Napoleonic era at the beginning of the nineteenth century, when this 'new' discipline experienced a boost in many parts of Italy.[126]

Graham was aware of how Brazil's plantations were supplanting an international market. For example, for the specific case of Bahia, he noted in her *Journal*:

> There are eighteen English mercantile houses established at Bahia, two French, and two German. The English trade is principally carried on with Liverpool, which supplies manufactured goods and salt, in exchange for sugars, rums, tobaccos, cottons, very little coffee, and molasses.... The province of Bahia, by its neglect of manufactures, is quite dependent on commerce.[127]

Humboldt's and Graham's Brazil was thus a highly anthropized environment, in which the varied flora that set the scene for travels and walks was not always spontaneous or even native, but often included a significant allochthonous component that rested on long-standing policies of strategic cultivation as well as transport and acclimatization of plant species sometimes from other continents.

Southeast Asia was also a destination for many travelers. An interesting case study is that of Thailand and Vietnam (Siam and Cochinchina, respectively). Competition among colonial powers to control Southeast Asia increased during the nineteenth century, and more and more Western-ers traveled to this region. Their trips were either for individual study pur-poses or were financed by the colonial powers to further their political and economic ambitions.[128]

Scottish naturalist George Finlayson (1790–1823) left interesting descrip-tions of nature in Thailand and Vietnam. He worked in Ceylon and Bengal as a British Army physician and surgeon in the late 1810s. In 1821–1822, as a medical officer, he accompanied the trade mission of the diplomat John Crawfurd (1783–1868) to the courts of Thailand and Vietnam, ordered by the governor-general of India. Finlayson returned to Calcutta in 1823, but died on the return trip to Scotland due to the adverse health impact of his travels. In spite of his short life, he gained scientific fame for his pioneering studies of the flora and fauna of southern Thailand and the Malay Penin-sula. The journal he kept during Crawfurd's mission was published posthu-mously in 1826 under the editorship of Sir Stamford Raffles, politician and fellow of the Royal Society. Titled *The Mission to Siam, and Hué the Capital of Cochin China*, the book gives us valuable descriptions of ecosystems in Southeast Asia.[129]

Sailing along the Thai coast, Finlayson was attracted by the shapes and colors of the local topography. In the 'mountain masses and rocky emi-nences', the predominant gray granite was mixed with less frequent red granite, veined here and there with syenite and 'perhaps primitive trap, masses of quartz, with schorl and talc embedded'. He also described the vegetation: palms grew wild, but there were also numerous other plant spe-cies accurately identified by the author with their scientific names. He emphasized the almost total lack of anthropogenic presence in the area: 'Thick, dense forest, without any trace of contiguous cultivation'.[130]

Other passages in *The Mission* contain detailed descriptions of nature on Ko Sichang Island and the nearby islands in the Gulf of Thailand, not only the flora—which Finlayson always described in detail—and geological aspects but also the fauna. While the variety of land mammals was rather small, there were numerous porpoises 'of a clear white color, with a very slight tinge of pink' swimming in the surrounding waters. Bird species were also numerous, such as hawks, pelicans, blue herons, and pigeons of various colors (bluish, iron-brown, green). Reptiles, crustaceans, and fish were also abundant.[131]

These excerpts present ecosystems teeming with life, in which the plant and animal kingdoms are presented to the eyes of the traveler and scholar in all their harmony. However, this was not always a nature devoid of anthropic presence, as Finlayson pointed out in other stretches of his journey. For example, on Ko Sichang Island, Finlayson reported that Crawfurd suddenly came upon a small plain in the jungle 'neatly cultivated with Indian corn, chillies, yams, and sweet potatoes'. It was a small plot admittedly poorly cared for by an elderly couple—he from China, she from Laos—shabby but serene in that simple life: 'They were exempt from many of the miseries that accompany a more civilized state'.[132] Elsewhere, however, Finlayson described the far more refined courts of Thailand and Vietnam, relations with which were, moreover, the main object of Crawfurd's mission.

In contrast, a far wilder and more hostile nature was witnessed by Elizabeth Bruce Elton Smith (1805–1854) in India. She traveled with her husband, Major Freeman Elton Smith, an army officer in the East India Company. From 1828, they lived in India and in 1832 returned to England because of Freeman's ill health. Upon their return, Elizabeth wrote *The East India Sketch-Book* (part one was published in 1832, part two in 1833), a collection of narrative sketches, memoires, and letters from Anglo-Indian society. It would be followed by a novel, *The Three Eras of Woman's Life* (1836). She later abandoned writing. She died in Schwetzingen, near Heidelberg, Germany.[133]

Elizabeth Bruce Elton Smith describes the Indian jungle as a gloomy, dangerous, and hostile environment when she had the opportunity to pass through it while traveling with her husband in the period between 1828 and 1832. Not surprisingly, she advised *Sketch-Book* readers who were unaware of what a jungle was that 'ignorance is bliss'.[134] Many of the descriptions she gave for that ecosystem were of this tenor:

> The jungle, as we advanced, became more dense; lofty hills environed us, covered with forests the abode of predatory animals, and that mightiest of serpents, the boa-constrictor. But how the terror of such foes faded beneath the dread of the pestilential vapours which were exhaling around us! Yes, unseasonable as it was, contrary to calculations of ordinary experience,—heavy rains deluged the earth, and threatened us with destruction. Morning after morning, our fearful eyes saw the heads of the encircling hills veiled in this black vapour, that was shortly to descend, and assail us as a pestilence.... Tall trees, or lofty forest-covered mountains, bounded our limited horizon, and seemed to shut in upon us the malaria abounding in the damp vegetation.[135]

The suffocating vegetation and the threat of predators combined with the damp and misty climate to arouse the fears of European travelers. However, again in the *Sketch-Book*, she admitted that in that 'mass of jungle' not everything was so unpleasant; she even waxed poetic in describing the fascinating aspects of the Indian jungle for 'the most world-tired man' should he choose to live isolated in those forests, surrounded by 'flowers dyed in the golden sunset' and 'insects innumerous' gamboling in the sun 'in glorious armour green and gold' or shining in the night like 'the stars of earth.'[136]

We close this section on travel in environments affected by colonialism with a case that somewhat postdates the others presented in this book but is interesting in its purpose. Here the colonial world is experienced by a European as a destination for 'health tourism'.[137] The Italian poet Guido Gozzano (1883–1916)—one of the leading exponents of the Italian literary current *Crepuscolarismo*—took a cruise in the Indian Ocean in 1912–1913, following the aggravation of a lung ailment diagnosed in 1907. He hoped that the warm, tropical marine climate would benefit his health, having had little success on the shores of the Mediterranean. Unfortunately, the 1912–1913 trip did not bring the hoped-for outcome, but it did inspire his writing. *Lettere dall'India* (Letters from India) was published in the Turin newspaper *La Stampa* in 1914 and later collected as a posthumous volume in 1917 titled *Verso la cuna del mondo* (Into the Cradle of the World).[138]

Moments of awe alternate with others of despondency and melancholy in Gozzano's work. For instance, homesickness is felt strongly in the telling of Christmas in Ceylon. Gozzano pictures the snow of his homeland while sweating in a tropical forest surrounded by a chorus of parrots and monkeys. From another perspective, tropical ecosystems inspire a sense of awe through the *Lettere*. Although agriculture is present and duly noted, the forest remains the defining element, appearing as a mysterious, attractive, yet dangerous realm.

> I step outside, refreshed from my bath, to distract myself as the forest awakens, a delight and wonder ever new to my European eyes. I follow a barely traced path in the density of the green, but for the first time this paradisiacal nature appears to me hostile, disturbing like an antediluvian landscape on which a *pleosauro* or an iguanodon might loom.... We advance on these narrow paths similar to corridors in the green, carved by the nocturnal excursions of wild elephants.[139]

Expressing a mix of amazement and discomfort not unlike that of Elizabeth Bruce Elton Smith eighty years earlier in India, Gozzano's point of view concludes this section, contrasting instead with the greater fascination experienced by Bory de Saint-Vincent on the islands in the Indian Ocean or by Maria Graham in Brazil.

We now shift to the analysis of a final type of tourism environment: one composed of reconstructed nature, contrasting strongly with the settings we have examined above and with those showing a significant anthropic presence, such as the 'tamed' nature of agricultural ecosystems and colonies.

2.8 A Final Perspective: Urban Parks and Gardens

Urban parks attracted the attention of tourists even before the twentieth century both for their natural aspects and for their sociocultural features. Among them we can include the botanical gardens of major European cities, which combined science and urban beauty and were often flaunted for the delight of influential visitors. Such was the case with the Di Negro Botanical Garden in Genoa. For example, in the summer of 1802, it was given a sunset visit to the Danish ambassador to Naples and his wife, who were passing through Genoa. Similarly, in 1805, the director of the Botanical Garden of the University of Bologna gave a tour to Marquis Spinola and a friend.[140] Moreover, thanks to the travel memoirs of the clergymen Giovanni Serafino Volta (1754–1842) and Luigi Marchelli (1837–1909), we have records of similar amenities in a society that was very different in customs and traditions from our own.

Giovanni Serafino Volta was a clergyman originally from the city of Mantua in northern Italy. He was a lifelong scholar of natural history and paleontology. He was one of the curators of the museum of natural history at the University of Pavia under the Habsburg government alongside the previously mentioned Lazzaro Spallanzani, his opponent in a number of scientific and personal diatribes.[141] Volta traveled through northern Italy and Mitteleuropa over the course of his life, studying the various natural and anthropic elements, from the mines in southern Hungary to the spa waters of Baden bei Wien, from the ecosystem of Lake Garda to the mountain regions between Austria and Italy. He left numerous handwritten or printed accounts of his travels.[142]

Volta has left us a number of pages on artificial nature in the city of Vienna and environs, which he was able to admire in first person.[143] He visited the gardens of Schönbrunn Palace with their rich variety of trees and plants collected over the decades and organized with supreme elegance. He stopped to dwell in particular on the bowers of horse chestnut and hornbeam, the Gloriette built on a panoramic rise in the park, and the greenhouses burgeoning with mimosa, palms, geraniums, epidendrums, various types of succulents, and other exotic species. The greenhouses were also home to numerous species of exotic birds—such as parrots in the genus *Psittacus*—which continued to reproduce 'as if they lived in their native land'. Volta also visited the botanical garden at the university, which contained both indigenous and allochthonous plant species for study purposes, although not able to boast the same variety and luxuriance of the gardens of Schönbrunn Palace. Both places featured the strong presence of the naturalist Nikolaus Joseph von Jacquin (1727–1817), who had traveled to Latin America in the 1750s at the behest of Francis I, Holy Roman Emperor, with the specific objective of collecting exotic plants to embellish the gardens of Schönbrunn Palace.[144]

Our second case study, Luigi Marchelli, was a Catholic priest from Pavia. His notebooks are rich in information not only about natural and human-made monuments and landscapes but also about the transformations and innovations occurring in Italian society in the 1860s and 1870s, during which Marchelli made numerous trips to northern and central Italy.[145]

Using contemporary definitions, we could say that Marchelli was a cultural tourist and an urban tourist, showing interest mainly in buildings, museums, churches, and exhibitions. He was an educated traveler, always buying guidebooks to his destinations. In addition to the built features of cities and towns, Marchelli also visited parks, especially public gardens and the parks of opulent villas, which were undergoing transformations and expansions in those very years or were being built from scratch on the model of the great English or French urban parks.

For instance, Marchelli visited the park of Villa Durazzo Pallavicini in the Ligurian town of Pegli (today part of the city of Genoa), giving a rich and enthusiastic description both of the vegetation and architecture. Once in Genoa, he visited the city's public gardens with his companions:

> Returning the way we came, we went up to Acquasola, which are the public gardens of Genoa. They are quite extensive and are embellished by two beautiful, very high jets of water, well aligned groves of orange trees, larch trees and flowers, and numerous shade trees, but greatly inferior to the Public Gardens of Milan.[146]

In the following years, Marchelli visited again the Acquasola Gardens, admiring changes and renovations. Nonetheless, he found beautiful urban

Table 2.1 Gardens and parks visited by Luigi Marchelli. Archivio Storico Diocesano di Pavia, VIII, 2, L. Marchelli, Viaggi – Taccuini. Our elaboration

City	Name
Milan	Old Public Gardens
	New Public Gardens
Turin	Valentino Gardens
	Balbo Gardens
Genoa	Acquasola Gardens:
	Old gardens
	New gardens
Pegli	Park of Villa Durazzo Pallavicini
Novara	The Allea (tree-lined boulevard)
Lodi	Public gardens
Mantua	Gardens of Palazzo Cavriani
	Piazza Virgiliana
Venice	Napoleonic public gardens
Padua	Prato della Valle
Florence	Public promenade to the Cascine
Modena	Ducal Garden
Parma	Park of the Ducal Palace
Piacenza	Public gardens

parks, gardens, and promenades in other parts of Italy too. For instance, he visited 'the public gardens near the Valentino' in Turin and strolled 'along the tree-lined boulevards around the town' in Novara (today's eastern Piedmont). In Venice, he visited the public gardens 'planted in 1807 by order of Napoleon'. In Padua, he was in the Prato della Valle. In Florence, he took walks to the Cascine following a public promenade located on the right bank of the Arno River and lined with venerable trees. In Lodi, not far from Milan, Marchelli made 'a pleasant visit to the new public gardens'. In Mantua, he strolled through the Piazza Virgiliana, 'all lined on the inside by a thick lane of trees with a path, and bordered by elegant residences'; he also admired the bronze bust of Roman poet Virgil (who had been born not far from Mantua). In Modena, he visited the 'Ducal Garden, now the Royal Garden … with an abundance of trees and flowers'. In Parma, he admired the Park of the Ducal Palace and in Piacenza he saw the small public gardens between the cathedral and the station.[147]

Marchelli visited the Public Gardens in Milan (today's Giardini Pubblici Indro Montanelli) many times over the years during his trips to that city. He found it an excellent place to rest, stroll, observe the natural beauty—such as the many varieties of animals and flowers—and visit industrial or artistic exhibitions in temporary pavilions located in the park. At the time, the Public Gardens were located near the former railway station, making them easily accessible by travelers coming from outside of Milan.[148]

Concluding our analysis, we can state that city parks and gardens are pillar ecosystems in the urban environment. Even the smallest green spaces in cities play an important role. The microbiomes that develop in urban parks support essential ecosystem services, such as atmospheric carbon dioxide sequestration or pollutant filtering. Urban green spaces also support microbiomes populated by microbes that are able to remove nitrogen from water or even amoebas that feed on potentially harmful bacteria. Urban parks are increasingly designed to contribute not only to the physical and mental well-being of nearby residents and city users but also to the environmental benefits of these social-ecological systems.[149]

Notes

1 Giuseppe Tagarelli and Francesco Torchia (eds), *Turismo, paesaggio e beni culturali. Prospettive di tutela, valorizzazione e sviluppo sostenibile* (Rome, 2021); Sandra Cavallo and Tessa Storey (eds), *Conserving Health in Early Modern Culture: Bodies and Environments in Italy and England* (Manchester, 2017); Raffaele Milani, *The Art of the Landscape* (Montreal and Kingston, 2009).
2 Carlo Cencini, *Il paesaggio naturale: i valori culturali della natura*, in Maria Chiara Zerbi (ed), *Il paesaggio rurale: un approccio patrimoniale* (Turin, 2007), pp. 27–45.

3 Paul Warde, Libby, Robin, Sverker Sörlin (eds), *The Environment: A History of the Idea* (Baltimore, 2018), pp. 25–35; Sharon E. Kingsland, *The Evolution of American Ecology, 1890–2000* (Baltimore, 2005); Pascal Acot (ed), *The European Origins of Scientific Ecology (1800–1901)* (Amsterdam, 1998).

4 Tiziano Tempesta, *La valutazione del paesaggio*, in Francesco Marangon (ed), *Gli interventi paesaggistico-ambientali nelle politiche regionali di sviluppo rurale* (Milan, 2006), pp. 58–76.

5 Eugenio Turri, *Il paesaggio come teatro. Dal territorio vissuto al territorio come rappresentato* (Venice,1998).

6 J. R. Brent Ritchie and Geoffrey I. Crouch, *The Competitive Destination: A Sustainable Tourism Perspective* (Wallingford and Cambridge MA, 2003).

7 John Robert McNeill, 'Observations on the Nature and Culture of Environmental History', *History and Theory* Volume 42, No. 4 (2003), pp. 5–43, specifically p. 6. For other definitions see: J. Donald Hughes, *What is Environmental History?* (Malden MA, 2006); Carolyn Merchant, *The Columbia Guide to American Environmental History* (New York, 2002); Donald Worster (ed), *The Ends of the Earth: Perspectives on Modern Environmental History* (Cambridge and New York, 1988).

8 Stephen Dovers, Ruth Edgecombe, Bill Guest (eds), *South Africa's Environmental History: Cases and Comparisons* (Athens OH, 2003).

9 Tempesta, *La valutazione del paesaggio*.

10 Enrico Banfi and Agnese Visconti, 'L'Orto di Brera alla fine della dominazione asburgica e durante l'età napoleonica', *Atti della Società Italiana di Scienze Naturali*, Volume II, No. 154 (2013), pp. 173–264; Giulio Sandri, *Elogio del dottor Ciro Pollini* (Verona, 1833). About the species of cochineal in question, see the uptodate research: Dalila Haouas, Lassaad Mdellel, I. Mraihi, C. Hafsi, V. Balmès, '*Pollinia pollini* (Costa, 1857) (Hemiptera, Asterolecaniidae) Infesting Olive Trees: A First Record in Tunisia', *Bulletin OEPP – EPPO Bulletin*, Volume 50, No. 1 (2020), pp. 201–202.

11 Ciro Pollini, *Viaggio al Lago di Garda e al Monte Baldo in cui si ragiona delle cose naturali di quei luoghi aggiuntovi un cenno sulle curiosita del Bolca e degli altri monti veronesi* (Verona, 1816), p. 4. Many references to Pollini's interest in the environment and agriculture in the Lake Garda area are present in the correspondence between him and Domenico Nocca, professor of botany at the University of Pavia, kept in BUPv, *Autografi*, 4, dossier Pollini.

12 Pollini, *Viaggio al Lago di Garda*, pp. 6–8.

13 *Goethe's Travels in Italy: toghether with His Second Residence in Rome and Fragments on Italy* (London, 1885), p. 20. See also Nikola Roßbach (ed), *Der See ging hoch mit seinen blauen, blauen, ach, so reizend blauen Wellen. Literatur zum Gardasee aus drei Jahrhunderten* (Vienna, 2014).

14 For Amoretti's biographical profile, his role in Italian scientific and cultural dissemination in the late eighteenth and early nineteenth century, and an overview of his traveling activity, see Agnese Visconti, 'Carlo Amoretti in viaggio tra Lombardia Austriaca e Mendrisiotto (1791): sentimenti d'amore e interessi scientifici', *Archivio Storico Ticinese*, No. 157 (2015), pp. 108–123; Franco Arato, 'Carlo Amoretti e il giornalismo scientifico nella Milano di fine Settecento', *Annali della Fondazione Luigi Einaudi*, Volume 21 (1987), pp. 175–216; Renzo De Felice, *Amoretti, Carlo*, in *Dizionario Biografico degli Italiani: volume 3* (Rome, 1961), https://www.treccani.it/enciclopedia/carlo-amoretti_%28Dizionario-Biografico%29/, accessed on December 2022.

15 For Arthur Young's biographical profile, see Gordon Edmund Mingay, *Young, Arthur*, in *Oxford Dictionary of National Biography* (Oxford, 2004), https://doi.org/10.1093/ref:odnb/30256, accessed on December 2, 2022.

16 Carlo Amoretti, *Viaggio da Milano a Nizza … ed altro da Berlino a Nizza e ritorno da Nizza a Berlino di Giangiorgio Sulzer fatto negli anni 1775 e 1776* (Milan, 1819), p. 3.

17 Giorgio Bigatti (ed), *Quando l'Europa ci ammirava: viaggiatori, artisti, tecnici e agronomi stranieri nell'Italia del '700 e '800* (Truccazzano, 2016); Christopher F. Black, *Early Modern Italy: a Social History* (London and New York, 2001), pp. 20–21, 39–40.

18 Amoretti, *Viaggio da Milano a Nizza*, p. 8.

19 Luciano Maffi and Martino Lorenzo Fagnani, *Disability and Tourism in Nineteenth- and Twentieth-Century Italy* (Abingdon and New York, 2021), pp. 17–78, 120–125; Alessandro Carassale, Daniela Gandolfi, Alberto Guglielmi Manzoni (eds), *Il viaggio in Riviera. Presenze straniere nel Ponente ligure dal XVI al XX secolo* (Bordighera, 2015); Andrea Zanini, *Un secolo di turismo in Liguria. Dinamiche, percorsi, attori* (Milan, 2012).

20 Amoretti, *Viaggio da Milano a Nizza*, pp. 19–23. About agriculture in Linguria see: Alessandro Carassale, 'Cedri e palme all'hebrea'. *Produzione e commercio nell'estremo Ponente ligure tra XVI e XVIII secolo*, in Alessandro Carassale, Claudio Littardi (eds), *Frontiera Judaica. Gli ebrei nello spazio ligure-provenzale dal medioevo alla Shoah* (Saluzzo, 2021), pp. 101–127; Alessandro Carassale and Claudio Littarti (eds), *Ars olearia: volume 2* (Guarene, 2019); Alessandro Carassale and Luca Lo Basso, *Sanremo, giardino di limoni. Produzione e commercio degli agrumi dell'estremo Ponente ligure (secoli XII-XIX)* (Rome, 2008).

21 Amoretti, *Viaggio da Milano a Nizza*, pp. 35, 45–49.

22 Carlo Amoretti, *Viaggio da Milano ai tre Laghi Maggiore, di Lugano e di Como, e ne' monti che li circondano* (Milan, 1794), p. 23. In a comparative way, it is interesting the study conducted on the 'lake poet' William Wordsworth and his perception of the English lakes as landscapes in Paul Smethurst, *Travel Writing and the Natural World, 1748–1840* (Basingstoke and New York, 2012), pp. 171–180.

23 Ibid., pp. 72–73.

24 Ibid., p. 38. Regarding Lombard sericulture, silk manufacturing and commerce see Angelo Moioli, *La gelsibachicoltura nella campagne lombarde dal Seicento alla prima metà dell'Ottocento* (Trento, 1981). For the Italian context see Giovanni Federico, *An Economic History of the Silk Industry, 1830–1930* (Cambridge and New York, 1997), pp. 112–117.

25 Both quotes are from Amoretti, *Viaggio da Milano ai tre Laghi*, p. 39.

26 ARJB, DIV. I, 20, 2, 5, letter by Alfonso Castiglioni to Professor Casimiro Gómez Ortega, Milan, January 15, 1785; BNB, AF XI 34, 117 verso, for Luigi Castiglioni's interest in the Seneca snakeroot.

27 BNB, AF XI 38: 3 recto, letter from Carlo Amoretti to the American Philosophical Society (called 'Società Economica'), March 1786; 3 verso, Amoretti to Alfonso Castiglioni, April 6, 1786; 18 recto, Amoretti to Alfonso Castiglioni, March 3, 1787.

28 Alfred W. Crosby, *The Columbian Exchange: Biological and Cultural Consequences of 1492* (Westport CT, 1972).

29 Rey Desmond, *Kew: The History of the Royal Botanic Gardens* (London, 2007, 2nd edition); Antonio González Bueno and Raúl Rodríguez Nozal, *Plantas americanas para la España ilustrada: génesis, desarrollo y ocaso del proyecto español de expediciones botánicas* (Madrid, 2000); Emma C. Spary, *Utopia's Garden:*

French Natural History from Old Regime to Revolution (Chicago and London, 2000); Lucile H. Brockway, *Science and Colonial Expansion: The Role of the British Royal Botanic Gardens* (New York and London, 1979).

30 For the testimony of Wollstonecraft Shelley, see Betty T. Bennet (ed), *Selected Letters of Mary Wollstonecraft Shelley* (Baltimore, 1995), pp. 296–300, Mary Shelley to Everina Wollstonecraft, Lake Como, July 20, [1840]. For the measures planned by authorities and scientific institutions for the relaunch of live growing in the Lake Como area in the eighteenth and nineteenth centuries, see ASMi, *Agricoltura p.a.*, 77, documentation from the 1770s to the 1790s; AIL, *Archivio storico*, section I, IV, 7, 3, documentation of 1819 and 1820. See also Agnese Visconti, *Paesaggi di Lombardia: il caso dell'ulivo tra ambienti naturali e tecniche manifatturiere (1772–1796)*, in Gabriella Guerci, Laura Pelissetti, Lionella Scazzosi (eds), *Oltre il giardino: le architetture vegetali e il paesaggio* (Florence, 2003), pp. 167–174.

31 *Hand-book for Travellers in Northern Italy* (London, 1847, 3rd edition), p. 123.

32 Ibid., p. 124

33 Mary Wollstonecraft Shelley, *Rambles in Germany and Italy in 1840, 1842 and 1843: volume I* (London, 1844), p. 76.

34 *Hand-book for Travellers in Northern Italy*, p. 124. For an interesting overview on women and work in the period in question, see Deborah Simonton, *A History of European Women's Work: 1700 to the Present* (London and New York, 1998).

35 *Hand-book for Travellers in Northern Italy*, p. 127

36 Wollstonecraft Shelley, *Rambles: volume I*, p. 99. For the socioeconomic context, see Silvia A. Conca Messina (ed.), *Leading the Economic Risorgimento: Lombardy in the 19th Century* (Abingdon and New York, 2022); Maurizio Romano, *Alle origini dell'industria lombarda. Manifatture, tecnologie e cultura economica nell'età della Restaurazione* (Milan, 2012).

37 Enrica Yvonne Dilk, *Un agronomo austriaco nel Lombardo-Veneto*, in Johann Burger, *Agricoltura del Regno Lombardo-Veneto: versione italiana del dottor V. P. con note del Dottor Giuseppe Moretti* [anastatic reprint of the 1843 Italian edition] (Milan, 2002), pp. 11–32.

38 Burger, *Agricoltura del Regno Lombardo-Veneto*, p. 131.

39 Dilk, *Un agronomo austriaco nel Lombardo-Veneto*, pp. 18–23.

40 About the spread of the British agricultural evolution in Europe in the eighteenth and part of the nineteeth centuries, see Peter M. Jones, *Agricultural Enlightenment: Knowledge, Technology, and Nature, 1750–1840* (Oxford and New York, 2016).

41 Burger, *Agricoltura del Regno Lombardo-Veneto*, pp. 15–21.

42 For instance, another important description of northern Italian agricultural system and in general the rural environment is provided in Thomas Jefferson, *Memorandums Taken on a Journey From Paris Into The Southern Parts of France and Northern Italy, in the Year 1787*, in Thomas Jefferson Randolph (ed), *Memoirs, Correspondence, and Private Papers of Thomas Jefferson, Late President of the United States: volume II* (London, 1829), pp. 115–160.

43 Patrizia Battilani, *Vacanze di pochi, vacanze di tutti. L'evoluzione del turismo europeo* (Bologna, 2007), pp. 120–136; Thomas Busset, Luigi Lorenzetti, Jon Mathieu (eds), *Tourisme et changements culturels – Tourismus und Kulturelle Wandel* (Zürich, 2004); Fergus Fleming, *Killing Dragons: The Conquest of the Alps* (New York, 2000).

44 Yannick Vialette, Pascal Mao, Fabien Bourlon, 'Scientific Tourism in the French Alps: A Laboratory for Scientific Mediation and Research', *Journal of Alpine Research – Revue de géographie alpine*, Volume 109, No. 2 (2021), https://doi.

org/10.4000/rga.9189, accessed on December 2, 2022; Margarida Archinard, 'Les instruments scientifiques d'Horace-Bénédict de Saussure', *Le Monde alpin et rhodanien. Revue régionale d'ethnologie*, Volume 1, No. 1–2 (1988), pp. 151–164; Andrew Beattie, *The Alps: A Cultural History* (Oxford and New York, 2006); Battilani, *Vacanze di pochi*, pp. 120–136.

45 Peter H. Hansen, *The Summits of Modern Man: Mountaineering after the Enlightenment* (Cambridge MA, 2013); René Sigrist, *La Nature à l'épreuve. Les débuts de l'expérimentation à Genève (1670–1790)* (Paris, 2011), pp. 359–381, 465–472, 503–533; Albert V. Carozzi, *Horace-Bénédict de Saussure (1740–1799): un pionnier des sciences de la terre* (Geneva, 2005); René Sigrist (ed), *H.-B. de Saussure (1740–1799): un regard sur la Terre* (Geneva, 2001).

46 Horace-Bénédict de Saussure, *Voyages dans les Alpes, précédés d'un essai sur l'histoire naturelle des environs de Genève: volume IV* (Geneva, 1786), p. 11.

47 Ibid., p. 405.

48 Amoretti, *Viaggio da Milano ai tre Laghi*, p. 23.

49 William Brockedon, *Illustrations of the Passes of the Alps, by which Italy Communicates with France, Switzerland, and Germany: volume II* (London, 1829), pp. 41–42. Regarding the Spluga Pass area see Kurt Wanner, *Lo Spluga. Il passo sublime* (Chiavenna, 2005).

50 Amoretti, *Viaggio da Milano ai tre Laghi*, pp. 67–68. On geological theories circulating in Italy at the turn of the nineteenth century see Gian Battista Vai, *Light and Shadow: The Status of Italian Geology Around 1807*, in Cherry L. E. Lewis, Simon J. Knell (eds), *The Making of the Geological Society of London* (London, 2009), pp. 179–202.

51 Leitch Ritchie and Clarkson Stanfield, *Travelling Sketches in the North of Italy, the Tyrol, and on the Rhine* (London, 1832).

52 Ibid., p. 73.

53 Marzio Achille Romani, *'Alpe' e 'Alpi'. Economie e società della montagna tra Medioevo e XIX secolo* (Brescia, 1987).

54 William A. J. Archbold and Elizabeth Baigent, *King, Samuel William*, in *Oxford Dictionary of National Biography* (Oxford, 2004), https://doi.org/10.1093/ref:odnb/15598, accessed on December 2, 2022.

55 Samuel William King, *The Italian Valleys of the Pennine Alps* (London, 1858), pp. 96–97.

56 Ibid., p. 101.

57 Josiah Gilbert and George Cheetham Churchill, *The Dolomite Mountains. Excursions through Tyrol, Carinthia, Carniola, and Friuli in 1861, 1862, and 1863* (London, 1864).

58 William Bainbridge, 'Titian Country: Josiah Gilbert (1814–1893) and the Dolomite Mountains', *Journal of Historical Geography*, Volume 56 (April 2017), pp. 22–42.

59 For instance, Gilbert and Churchill, *The Dolomite Mountains*, pp. xviii–xix, 379.

60 Ibid., p. xv.

61 Ibid., p. xvii.

62 Official Website of the Dolomites UNESCO World Heritage, https://www.dolomitiunesco.info/, accessed on December 2, 2022. See also Bruno Zanon (ed), *Le Dolomiti. Patrimonio mondiale UNESCO. Fenomeni geologici e paesaggi umani* (Pisa, 2021).

63 Cesare Micheletti, *Landscape value*, in Official Website of the Dolomites UNESCO World Heritage, https://www.dolomitiunesco.info/universal-values-dolomites-unesco/il-valore-del-paesaggio/?lang=en, accessed on December 2, 2022.

64 Gilbert and Churchill, *The Dolomite Mountains*, p. 9.

65 Ibid., pp. 30–38, while the quote is from p. 45.

66 Ibid., pp. 70–73.

67 For a biographical profile of Conway and some information about his family background, see Peter H. Hansen, *Conway, (William) Martin, Baron Conway of Allington*, in *Oxford Dictionary of National Biography* (Oxford, 2004), https://doi.org/10.1093/ref:odnb/32536, accessed on December 2, 2022; Joan Evans, *The Conways: A History of Three Generations* (London, 1966).

68 William Martin Conway, *The Alps from End to End* (Westminster, 1895), p. 107.

69 Ibid., p. 94.

70 Matti Seppälä (ed), *The Physical Geography of Fennoscandia* (Oxford and New York, 2005). See also Nina Wormbs (ed), *Competing Arctic Futures: Historical and Contemporary Perspectives* (Cham, 2018).

71 Roberto Navarrini, *Cenni biografici di Giuseppe Acerbi*, in Roberto Navarrini (ed), *Le Carte Acerbi nella Biblioteca Teresiana di Mantova. Inventario* (Rome, 2002), pp. vii–xiii. Skjöldebrand also recorded a description of this experience: Anders Friedrik Skjöldebrand, *Voyage pittoresque au Cap Nord* (Stockholm 1801–1802).

72 Giuseppe Acerbi, *Travels through Sweden, Finland, and Lapland, to the North Cape, in the Years 1798 and 1799: volume I* (London, 1802), p. 182. Also see these recent editions: Giuseppe Acerbi, *Il viaggio in Lapponia (1799)*, Luigi G. De Anna and Lauri Lindgren (eds) (Turku, 2009, 2nd edition); Giuseppe Acerbi, *Il viaggio in Svezia e in Finlandia (1798–1799)*, Lauri Lindgren (ed) (Turku, 2005); Giuseppe Acerbi, *Il viaggio in Svezia e in Norvegia (1799–1800)*, Lauri Lindgren and Luigi G. De Anna (eds) (Turku, 2000). See also Vincenzo De Caprio, 'Sul paesaggio finlandese e l'idea di natura in Giuseppe Acerbi', *Atti e memorie – Accademia Nazionale Virgiliana di Scienze Lettere e Arti*, Volume 85 (2017), pp. 185–206.

73 Acerbi, *Travels: Volume I*, pp. 183–185.

74 Ibid., pp. 187–188

75 De Caprio, 'Sul paesaggio finlandese e l'idea di natura in Giuseppe Acerbi', pp. 203–205.

76 Valerio Giacomini, *Brocchi, Giovanni Battista*, in *Dizionario Biografico degli Italiani: Volume 14* (Roma, 1972), https://www.treccani.it/enciclopedia/giovanni-battista-brocchi_%28Dizionario-Biografico%29/, accessed on December 4, 2022.

77 Chauncey C. Loomis, *Weird and Tragic Shores: The Story of Charles Francis Hall* (New York, 2000).

78 Andrew Lambert, *Franklin: Tragic Hero of Polar Navigation* (London, 2009); Clive Holland, *Franklin, Sir John*, in Francess G. Halpenny (ed), *Dictionary of Canadian Biography: volume 7* (Toronto, 1988), pp. 323–328. About the finding of the two wrecks, see, for instance, the dedicated webpage of the Royal Museums Greenwhich, https://www.rmg.co.uk/stories/topics/hms-terror-erebus-history-franklin-lost-expedition, accessed on December 4, 2022.

79 For an in-depth analysis of Hall's expedition and his perspective on Inuit culture, see Karen Routledge, *Do You See Ice? Inuit and American at Home and Away* (Chicago and London, 2018); Hester Blum, *Charles Francis Hall's Arctic researches*, in Steve Mentz, Martha Elena Rojas (eds), *The Sea and Nineteenth-Century Anglophone Literary Culture* (Abingdon and New York, 2017), pp. 47–65.

80 Charles Francis Hall, *Arctic Researches and Life among the Esquimaux: Being the Narrative of an Expedition in Search of Sir John Franklin, in the Years 1860, 1861, and 1862* (New York, 1865), p. 35.

81 Ibid., pp. 36–38, 83–85. The quote is from p. 85.
82 Hall, *Arctic Researches*, pp. 171, 194.
83 Ibid., p. 151.
84 King, *The Italian Valleys*, pp. 206–207.
85 Ibid., p. 207.
86 Ibid., p. 207.
87 Ibid., p. 288.
88 Ibid., p. 290.
89 For an interesting survey of desert ecosystems, refer to Julie J. Laity, *Deserts and Desert Environments* (New York, 2009).
90 Lucien Neumann, *Biographies vétérinaires* (Paris, 1896), p. 80; Louis Damoiseau, *Voyage en Syrie et dans le désert* (Paris, 1833).
91 Damoiseau, *Voyage en Syrie*, pp. 67–70.
92 Ibid., pp. 70–71.
93 Ibid., p. 71. The quote is from the page *Ancient Jericho / Tell es-Sultan*, in UNESCO website, https://whc.unesco.org/en/tentativelists/6545/, accessed on December 4, 2022.
94 Damoiseau, *Voyage en Syrie*, pp. 71–72.
95 Richard Garnett and Elizabeth Baigent, *Browne, William George*, in *Oxford Dictionary of National Biography* (Oxford, 2004), https://doi.org/10.1093/ref:odnb/3710, accessed on December 5, 2022.
96 William George Browne, *Travels in Africa, Egypt, and Syria, from the Year 1792 to 1798* (London, 1799), pp. 26–27.
97 Ibid., pp. 184–187.
98 Tilar J. Mazzeo, *Volney, Constantin François de Chasseboeuf, Comte de*, in *Encyclopedia of the Romantic Era 1760–1850* (New York and London, 2004), pp. 1195–1196.
99 Constantin-François Volney, *Voyage en Syrie et en Egypte, pendant les années 1783, 1784 et 1785: volume I* (Paris, 1787, 2nd edition), pp. 55–58.
100 Giovanni Battista Brocchi, *Giornale delle osservazioni fatte ne' viaggi in Egitto, nella Siria e nella Nubia: volume I* (Bassano del Grappa, 1841), pp. 59–60, 220–223, 358–359.
101 Brian Austin, *Marine Microbiology* (Cambridge and New York, 1988), pp. 8–11. More in general refer to Alan Longhurst, *Ecological Geography of the Sea* (Burlington, San Diego, and London, 2007, 2nd edition).
102 Andreas W. Daum, *German Naturalists in the Pacific around 1800: Entanglement, Autonomy, and a Transnational Culture of Expertise*, in Hartmut Berghoff, Frank Biess, Ulrike Strasser (eds), *Explorations and Entanglements: Germans in Pacific Worlds from the Early Modern Period to World War I* (New York, 2019), pp. 70–102; Roderick J. Barman, 'The Forgotten Journey: Georg Heinrich Langsdorff and the Russian Imperial Scientific Expedition to Brazil, 1821–1829', *Terrae Incognitae*, Volume 3, No. 1 (1971), pp. 67–96. See also Anne S. Troelstra, *Bibliography of Natural History Travel Narratives* (Leiden, 2016).
103 George Heinrich von Langsdorff, *Voyages and Travels in Various Parts of the World, during the Years 1803, 1804, 1805, 1806, and 1807* (London, 1817), p. 23. Both quotes are from this page.
104 Ibid., p. 35.
105 For examples, see Langsdorff, *Voyages and Travels*, pp. 79–80.
106 Ibid., p. 80.
107 Ibid., pp. 419–420.
108 Ibid., pp. 421–422.
109 Ibid., p. 422.

110 Regina Akel, *Maria Graham: A Literary Biography* (Amherst and New York, 2009). About the improtance of traveling in Graham's life, refer mainly to: Claudia Georgi, *Maria Graham, travel writing on India, Italy, Brazil, and Chile (1812–1824)*, in Barbara Schaff (ed), *Handbook of British Travel Writing* (Berlin and Boston, 2020), pp. 313–334; M. Soledad Caballero, *Clashing tastes: European femininity and race in Maria Graham's journal of a voyage to Brazil*, in edited by Miguel A. Cabañas, Jeanne Dubino, Veronica Salles-Reese, Gary Totten (eds), *Politics, Identity, and Mobility in Travel Writing* (Abingdon and New York, 2015).

111 Maria Graham, *Journal of a Voyage to Brazil and Residence There During Part of the Years 1821, 1822, 1823* (London, 1824), p. 90.

112 Ibid., pp. 203–204.

113 Ibid., p. 203.

114 Alfred W. Crosby, *Ecological Imperialism: The Biological Expansion of Europe, 900–1900* (Cambridge and New York, 2004, 2nd edition).

115 Helen Tilley, *Africa as a Living Laboratory. Empire, Development, and the Problem of Scientific Knowledge, 1870–1950* (Chicago and London, 2011).

116 Corey Ross, *Ecology and Power in the Age of Empire: Europe and the Transformation of the Tropical World* (Oxford, 2017); Londa Schiebinger and Claudia Swan, *Colonial Botany: Science, Commerce, and Politics in the Early Modern World* (Philadelphia, 2005); John F. Richards, *The Unending Frontier: An Environmental History of the Early Modern World* (Berkeley, Los Angeles, and London, 2003), pp. 366–372.

117 André Role, *Un destin hors série : la vie aventureuse d'un savant : Bory de Saint-Vincent 1778–1846* (Paris, 1973).

118 *Le Morne Cultural Landscape*, in UNESCO website, https://whc.unesco.org/en/list/1259/, accessed on December 12, 2002. See also Le Morne Heritage Trust Fund website, https://lemorneheritage.org/biodiversity.html, accessed on December 12, 2022.

119 Jean Baptiste Bory de Saint-Vincent, *Voyage to, and Travels through the Four Principal Islands of the African Seas: Performed by Order of the French Government, during the Years 1801 and 1802: volume II* (London, 1805), pp. 71–72.

120 Ibid., p. 96.

121 See Chapter 1.

122 Graham, *Journal of a Voyage to Brazil*, p. 116.

123 Ibid., p. 143. See also Robert Southey, *History of Brazil: Volume III* (London, 1819), p. 797.

124 Julia Bruce, 'Banks and Breadfruit', *RSA Journal*, Volume 141, No. 5444 (1993), pp. 817–820.

125 Hippolyte Nectoux, *Observations sur la préparation des envois de plantes et arbres des Indes Orientales pour l'Amérique, et leur traitement pendant la traversée*, in *Mémoires d'agriculture, d'économie rurale et domestique publiés par la Société royale d'agriculture*, winter issue (1791), pp. 110–123. See also Francesca Bray, Barbara Hahn, John Bosco Lourdusamy and Tiago Saraiva, Moving Crops and the Scales of History (New Haven, 2023), pp. 113–118; Emma C. Spary, *Eating the Enlightenment: Food and the Sciences in Paris* (Chicago and London, 2012), pp. 84–89; Spary, *Utopia's Garden*, pp. 90–91.

126 A good example is given by the correspondence and related plant species interchanges between the agriculturists of the University of Pavia and the Paris Botanical Garden regarding both cereal cultivation improvements and potential exotic crops: in BCMHN, see the letters written by Professor Giuseppe

Bayle Barelle (Ms 1971, 152–153) and Professor Giovanni Biroli (Ms THO 367/1 and Ms THO 383/2) to Professor André Thouin.

127 Graham, *Journal of a Voyage to Brazil*, p. 146.

128 About traveling experience in East and Southeast Asia through history, an interesting and multifaceted point of view is given in Steve Clark and Paul Smethurst (eds), *Asian Crossings: Travel Writing on China, Japan and Southeast Asia* (Aberdeen HK, 2008).

129 Stamford Raffles (ed), *The Mission to Siam, and Hué the Capital of Cochin China, in the Years 1821–1822, from the Journal of the Late George Finlayson, Esq.* (London, 1826), pp. vii–xxvi. See also Anne S. Troelstra, *Bibliography of Natural History Travel Narratives* (Leiden, 2016), p. 153.

130 Raffles (ed), *The Mission to Siam*, pp. 10–11.

131 Ibid., pp. 273–277.

132 Ibid., pp. 277–278.

133 For a biography of Elizabeth Bruce Elton Smith, see Rosemary Raza, *In Their Own Words: British Women Writers and India 1740–1857* (Oxford, 2006), p. 273.

134 Elizabeth Bruce Elton Smith, *The East India Sketch-Book: Comprising an Account of the Present State of Society in Calcutta, Bombay, etc.: Volume 2* (London, 1832), p. 113.

135 Ibid., pp. 42–43.

136 Ibid., pp. 117–118.

137 On the concept of health tourism see, for instance, Maffi and Fagnani, *Disability and Tourism*, pp. 17–78.

138 For a biographical profile of Guido Gozzano, see Marziano Guglielminetti, *Gozzano, Guido*, in *Dizionario Biografico degli Italiani: Volume 58* (Rome, 2002), https://www.treccani.it/enciclopedia/guido-gozzano_%28Dizionario-Biografico %29/, accessed on December 6, 2022. See also Mariarosa Masoero, *Guido Gozzano, Libri e lettere* (Florence, 2005), and the rich bibliography referred to in Masoero's book.

139 Guido Gozzano, *Verso la cuna del mondo: Lettere dall'India (1912–1913)* (Milan, 1917), p. 34. For the analysis in this chapter, we refer to the 1998 edition published by E.D.T., Turin.

140 ADG, *Doria di Montaldeo*, 576/1011A, August 9, 1802, and 564/978A, September 19, 1805.

141 Fulvio Baraldi, 'Giovanni Serafino Volta, chimico, mineralogista e paleontologo mantovano (Mantova, 1754–1842)', *Atti e memorie – Accademia Nazionale Virgiliana di Scienze Lettere e Arti*, Volume 81 (2013), pp. 17–46 and the rich bibliography referred to in the essay.

142 BUPv, *Aldini*, 531, two handwritten volumes titled 'Saggi di storia naturale, chimica e fisica di Giovanni Serafino Volta' collecting scientific and travel essays by Volta. For the description he made of the Lake Garda area, see Giovanni Serafino Volta, *Descrizione del Lago di Garda e de' suoi contorni con osservazioni di storia naturale e di belle arti* (Mantua, 1828).

143 BUPv, *Aldini*, 531, pp. 165–180: the report is titled 'Osservazioni sui giardini, gabinetti di storia naturale, ed altre particolarità della capitale di Vienna' meaning 'Observations on gardens, cabinets of natural history, and other peculiarities of the capital Vienna'.

144 For the description of the gardens of Schönbrunn Palace and the University Botanic Garden, see BUPv, *Aldini*, 531, pp. 165, 172–173. About Jacquin refer mainly to Santiago Madriñan, *Nikolaus Joseph Jacquin's American Plants Botanical Expedition to the Caribbean (1754–1759) and the Publication of the Selectarum Stirpium Americanarum Historia* (Leiden and Boston, 2013).

145 ASDPv, VIII, 2, L. Marchelli, Viaggi – Taccuini.
146 ASDPv, VIII, 2, L. Marchelli, Viaggi – Taccuini.
147 ASDPv, VIII, 2, L. Marchelli, Viaggi – Taccuini.
148 ASDPv, VIII, 2, L. Marchelli, Viaggi – Taccuini.
149 Joana Vieira, Paula Matos, Teresa Mexia, Patrícia Silva, Nuno Lopes, Catarina Freitas, Otília Correia, Margarida Santos-Reis, Cristina Branquinho, Pedro Pinho, 'Green spaces are not all the same for the provision of air purification and climate regulation services: The case of urban parks', and Teresa Mexia, Joana Vieira, Adriana Príncipe, Andreia Anjos, Patrícia Silva, Nuno Lopes, Catarina Freitas, Margarida Santos-Reis, Otília Correia, Cristina Branquinho, Pedro Pinho, 'Ecosystem services: urban parks under a magnifying glass', both in *Environmental Research*, Volume 160 (January 2018), respectively pp. 306–313 and pp. 469–478.

3 Travel Arrangements

3.1 What Are We Analyzing?

Tourist travel rests on careful arrangements made by the traveler prior to departure and the infrastructure/services available at the destination. While this is true of all forms of tourism, even the long weekend at a vacation farm in one's own country, there is a greater feeling of consequentiality when the destination is in a distant land with a relatively scarce human presence. Currently, tourism in natural ecosystems ideally takes the form of responsible travel to areas that have measures in place to safeguard ecosystem integrity and produce economic benefits for local communities, thus further encouraging conservation. At the borderline between the anthropic and the natural, this sort of environmental tourism combines the goals of ecological conservation and economic development. Unfortunately, tourism in natural ecosystems can also be coupled with unsustainable development that causes or exacerbates environmental deterioration. To ensure that this type of tourism delivers on its promises, policymakers, economic actors, and tourists work toward ensuring that all tourism-related activities have a minimal impact on the environment and on the activities and culture of the people living in those areas.[1]

In this chapter, we have selected three fundamental aspects of a trip to a foreign land, focusing mainly, but not exclusively, on a wilderness context: transportation, food and accommodation, and contact with local people. These are still critical elements in the planning and management of a tourist trip to wilderness or natural areas (and also in scientific expeditions), and we may imagine that they were also critical in travels to similar contexts in the eighteenth and nineteenth centuries. We analyze the impressions recorded by travelers regarding organizational aspects and the difficulties encountered and how these were overcome.

DOI: 10.4324/9781003230519-4

3.2 Transportation

Tourism in natural or agricultural ecosystems can theoretically contribute to local development in a sustainable way by creating attractive sources of livelihood that are mutualistically interconnected with the conservation and/or enhancement of biodiversity. It is well established that the preservation of biodiversity is critical to preserving the significant economic benefits that countries derive from tourism. But how, for example, can biodiversity conservation and infrastructure development be reconciled to enable host countries to welcome tourists? Despite the current importance and awareness of this question, the degree to which biodiversity drives tourism patterns, especially in relation to infrastructure, is little known. A recent study analyzing birdwatching in natural ecosystems in Costa Rica, for example, suggests that infrastructure investments must be coupled with effective biodiversity conservation in order for tourism to generate economic revenues that protect the environment and its ecosystems.[2]

Analysis of the documentation reveals a close dependency of tourism on adequate local infrastructure throughout the course of the eighteenth century and much of the nineteenth century. In almost all recorded natural destinations, the road network was not extensively developed. A more extensive transportation system appears in the second half of the nineteenth century in much of Europe and the more urbanized areas of the world. We offer a few examples, both taking travel experiences from previous chapters and adding new ones.

We begin Section 3.2 by considering river transport, then move on to sea transport, and finally land transport. The first case from which we can draw interesting information about river transportation is that of Cornelis De Bruijn (1652–1726 or 1727), a Dutch painter who traveled to different parts of Europe, Egypt, and Asia. His experiences as a traveler can be grouped into two long journeys, the first stretching from 1674 to 1693 and the second from 1701 to 1708. We focus on the latter, which was motivated by his desire to visit the ruins of Persepolis, one of the ancient capitals of the Persian Achaemenid Empire and now a UNESCO World Heritage Site in Iran. In truth, the trip was rich in other experiences as well. For example, having visited Persepolis, he went as far as the Dutch East Indies and the capital of Indonesia, Jakarta (then known as Batavia). During the first part of the journey, he visited Russia and had the opportunity to stay a few months at the court of Peter the Great in Moscow, where he met the tsar personally and performed a number of artistic commissions for him.[3]

Let us focus precisely on the Russian interlude. De Bruijn made a trip on the Volga in 1703. From the morning of April 24 to the evening of May 20, he sailed from a river port near Moscow to Astrakhan on the Caspian Sea. His companion was Jacob Daviedof, a merchant from Esfahan who had been in Holland and had lived for some time in Amsterdam. This voyage allows us to analyze some interesting initial information about river transportation. De Bruijn embarked together with fifty-one other Russian and 'Armenian' passengers and twenty-three sailors on a river boat. He made a drawing of the boat and gave a very detailed description of it:

> In these parts they have small flat-bottomed vessels, ... which carry about 300 bales of silk, or about 15 lasts: they are capacious, and have but one mast and one sail, a very large one, and of use chiefly when the wind is aft; but when the wind is not either right astern, or well upon the quarters, they row with sixteen oars. They have no rudder, but a long kind of paddle, broad at the end in the water, the other end is supported by a kind of crutch adapted to the purpose....[4]

The voyage was relatively smooth and with few mishaps. De Bruijn describes a peaceful journey in a rural setting characterized by frequent manmade elements. The landscape passing before his eyes was a sequence of forests, highlands, cultivated fields, and grazing herds alternating with villages, monasteries, and larger towns, such as Nizhny Novgorod where the Oka River flows into the Volga, or the imposing Kazan with its fortified citadel and many religious buildings.[5]

While other parts of De Bruijn's journey were not so smooth, as we will see in a few pages, the navigation on the Volga was characterized by many comforts. He counted servants among the total of fifty-two passengers, thus alluding to social distinctions but especially a division of roles on the boat, and a crew of twenty-three men. Passengers enjoyed food replenished at stops along the way or obtained directly from local fishermen in exchange for other goods, such as brandy (which was thus present and enjoyed on board). There were also extra services, such as drinks cooled with ice, which on at least one occasion, near Kazan, 'our people' went ashore to fetch. Once again it was clear that, assuming that at least some of the provisions were shared by all the passengers, often only some of them were instructed to go ashore for purchases.[6]

Considering these dynamics, it is difficult to align De Bruijn's voyage on the Volga in the spring of 1703 with today's experience of nature tourism, but neither is it rural tourism or the pescatourism we hypothesized for Alberto Fortis in Dalmatia in Chapter 1. Today, cruises are organized along the entire Volga to Astrakan, with stops along the way in major cities and historical sites. Between them the cruise passes through agricultural or forested land. The presence of domesticated nature in both De Bruijn's voyage and today's cruises therefore places the Volga experience primarily in a leisure tourism context. Moreover, the comfort in which De Bruijn traveled was due to close acquaintances in Moscow, in the court of Peter the Great, as well as contacts with Dutch notables residing in the city, so it rested on a solid network of prestigious and international acquaintances that was not within everyone's reach.[7]

A completely different setting characterized Giuseppe Acerbi's description of his canoe ascent of the Kattilakoski rapids on the Tornio River, along the border between the present-day states of Sweden and Finland near the Arctic Circle, 'a long series of water-falls, formed by the stony bed of the river, and by huge rocks which rise above the surface of the water'. Between 1798 and 1799, the Italian nobleman Acerbi, a future writer and diplomat but at the time still a carefree twenty-five-year-old traveler, crossed the Scandinavian peninsula together with a Swedish exploratory expedition and reached the North Cape. The ascent of the Tornio was a preferable alternative to the difficult overland journey, which was almost impossible in that area because of the tangled undergrowth and thick pine and fir forests as well as the marshy terrain 'where you are in danger every step of sinking in the mire'.[8]

As noted earlier, the two volumes in English published in London as *Travels through Sweden, Finland, and Lapland, to the North Cape*, which described his adventures, date to 1802. In his book, Acerbi observed how the rushing cataracts of the Kattilakoski rapids were particularly arduous to ascend, but the 'Finlandish Laplanders' who led the expedition's canoes knew perfectly well how to maneuver through them. In pairs, they would position themselves at either end of the canoe, use a long pole to assay the river bottom, 'find their point of resistance', plant the pole with all their strength, and push the boat upstream. The pole had precise characteristics: it was made of pine wood and about fifteen feet long. Acerbi concluded that 'it is a Herculean labor' and emphasized that a specialized crew was needed, one that could react to the endless obstacles that might be encountered during such arduous navigation. At the most critical point in the rapids, Acerbi and the other travelers disembarked and continued on foot for a long stretch along the riverbank so that the boatsmen could maneuver their craft with greater agility.[9]

For example, the boatman might accidentally plant the pole, not firmly into the riverbed, but on a smooth stone, with the result that it would slip as soon as he put his weight to it. He would react immediately and try to find a secure brace point in time, but the strong current of the Kattilakoski rapids could swiftly rotate the canoe, with its bow downstream or crosswise to the current, with the great risk of capsizing. Experienced Lappish boatmen had to be able to handle these emergency situations and also be familiar with stretches where the bank was passable and passengers could proceed on foot to avoid the dangers of the river.[10]

The navigation system on the Tornio was organized into stages with boats and crew being changed at each town. Acerbi pointed out that the Tornio

Figure 3.1 The Kattilakoski Rapids on the Tornio River. Giuseppe Acerbi, *Vues de la Suede, de la Finlande, et de la Lapponie, depuis le detroit du Sund jusq'au Cap-Nord* (Paris, 1803), plate 11.

was dangerous along its entire course—and not just in the portion with the Kattilakoski rapids—with characteristics varying in each stretch. It was thus necessary to have boatmen who were thoroughly familiar with each segment of the route. Local expertise was thus at a premium: 'it is but prudent to change boat and boatmen at each village, as the peasants are all perfectly masters of the channel in their respective bounds'.[11]

We can assume that, given the wilderness nature of the area described by Acerbi, not many foreign travelers navigated the Tornio. Knowledge of the river and the surrounding area, with forms of expertise developed for each area and alternative routes through the forests, thus met local practical needs for travel and trade. However, today the Tornio and the Kattilakoski rapids are popular destinations for both nature tourism and sport and experiential tourism, with picnic areas, traditional lean-tos (*laavu*), camping, and log rafting competitions. One can also go rafting with appropriate watercraft and guides here as in other areas of Finland, one example being the Ruuna Hiking Area near the Russian border.[12]

Acerbi showed an interest in local navigation methods also on other occasions. He sought information on the work of a priest in the coastal city of Kalajoki who had supervised the construction of a 'birch-bark boat' that could hold four people and was so light it could be transported by one person over land. The clergyman had shared his design with the Academy of Sciences and there were various cases in various regions of Svenskfinland where the locals implemented initiatives to improve land transportation.[13] Acerbi, however, was not the only traveler who, using local boats to explore a semi-wild territory, was an unintentional precursor to what today might be called 'alternative' tourism.

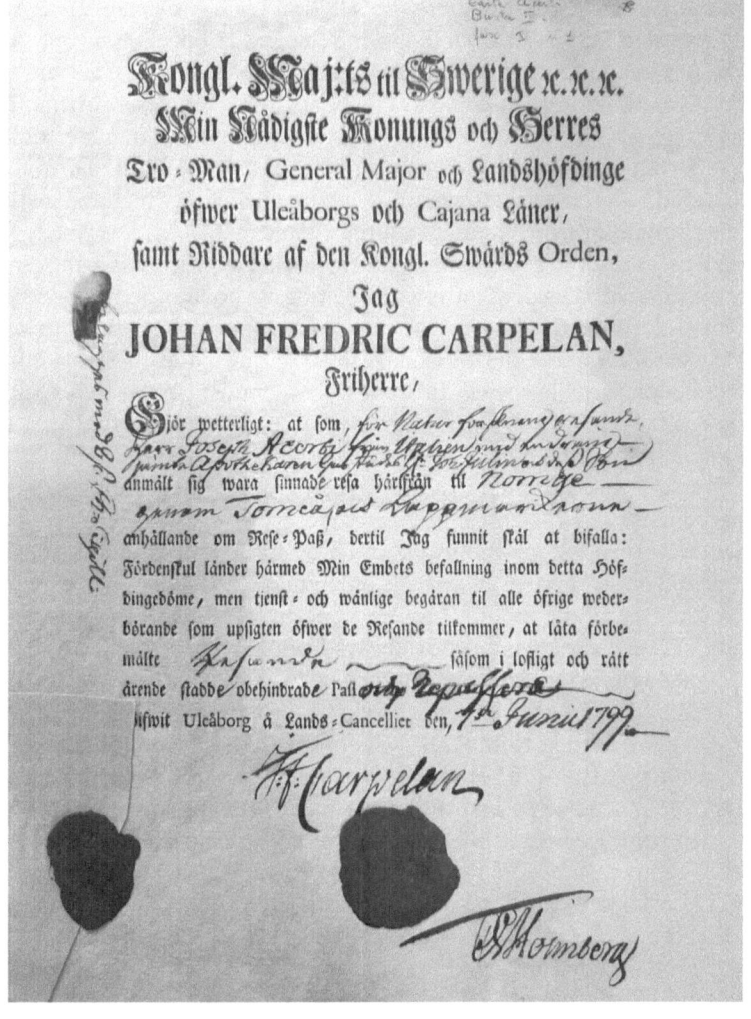

Figure 3.2 Passport issued to Giuseppe Acerbi by the Swedish authorities, Uleåborg, June 7, 1799. Biblioteca Comunale Teresiana (Mantua, Italy), *Fondo Giuseppe Acerbi, Carte*, busta III, fasc. I, doc. no. 1/1.

The geologist Giovanni Battista Brocchi, whom we have had the opportunity to analyze in depth in other contexts, also reports interesting information about water transportation during his travels, which were not always smooth. For example, when exploring the Egyptian hinterland in search of mines for Mehmet Ali in 1822–1823, Brocchi and the other members of the expedition sailed up the Nile from Cairo to Qena on feluccas, boats that are still widely used today in Egypt and some Middle Eastern states, also employed in a tourist context for 'alternative cruises' on the Nile.[14]

His *Giornale delle osservazioni fatte ne' viaggi in Egitto, nella Siria e nella Nubia* reported that three double-lateen feluccas had been assigned to him and the rest of the company, which consisted of colleagues, guides and interpreters, guards, and twelve miners. The boats were completely open, not an ideal design for winter traveling, which is when Brocchi undertook his voyage. To cope with the low temperatures, especially at night, half the boat was covered with straw mats, creating a 'sort of hut' in which the geologist and his companions sheltered as best they could.[15]

Needless to say, the geologist was far from satisfied with this arrangement: the scant protection against the low temperature—he recorded a temperature of 5°Ré (6.25°C) at dawn—and the fact that he was with seven companions plus 'the servants' under the same makeshift canopy made it, at least initially, a rather uncomfortable voyage. Brocchi and other members of the expedition had asked the authorities for a larger vessel, but they were told that none were available, that the Nile in those months was not high enough for a large-hulled vessel, and that, given that they were not of dignitary rank, they would have to make do. Brocchi's initial discontent was also exacerbated by his departure against the wind, hastily urged by the Egyptian authorities who had commissioned him 'as if we were off to find the Golden Fleece'.[16]

It is true that in the month it took to sail up from Cairo to Qena, sailing conditions were variable and not always unfavorable, with occasional fair winds and daytime temperatures. Brocchi recorded January 14 as 'a brilliant day', with a 'pleasant northerly breeze' that promised good sailing (were it not for repairs needed to the hull, which was damaged by rubbing against sandbars at night) and 8°Ré (10°C) as early as seven in the morning. On January 20, Brocchi even had the opportunity to complain about a sudden rise in temperature caused by the 'hot and stifling' sirocco that blew night and day. Temperatures rose from 13°Ré (16.25°C) at seven in the morning to 22°Ré (27.5°C) between noon and two in the afternoon.[17]

Navigation on the Nile also allowed Brocchi to see objects of special scientific interest that he could freely admire from his floating observatory. A sign that felucca navigation, with few barriers to shelter him but also few to block the view, also had its advantages.

One evening, the sky generally obscured by a strange fog, a completely clear patch hovered over him for several hours, affording a beautiful view of the stars and a quarter moon. In astonishment, Brocchi noted that the perfectly circular portion of cleared sky looked as if it had been drawn with a compass, and the diameter appeared to the eye to be one half larger than that of the dome of the Pantheon in Rome seen from the inside. A few days later, however, the nearly full moon was obscured by a total eclipse that Brocchi and his companions had not anticipated and that prevented them from sailing for at least two hours, as the tramontana was blowing hard and it was better not to proceed uncertainly. What might have been seen as a setback to progress up the river instead gave Brocchi the chance to witness the curious spectacle of the inhabitants coming out of their homes on the shores and banging frying pans and other metal objects or firing their rifles: the eclipse was considered a harbinger of grim news. Other sources of excitement were the crocodiles he saw for the first time basking in the afternoon sun on the sandy banks past Gerga (Thinis).[18]

The last example of river travel we propose is that of Mary Kingsley (1862–1900), an English ethnologist who made two trips to Atlantic Africa. In travels taken between 1893 and 1895, she visited various lands: Cabinda, Nigeria, Equatorial Guinea, French Congo, Gabon, and Cameroon. In 1900, during the Anglo-Boer War, she went down to Cape Town as a volunteer and was assigned to assist Boer prisoners of war. After a few months, she died of typhoid fever. Having lived in contact with African peoples and European missionaries, she had collected a great deal of material on the cultures of Atlantic Africa, their relations with the Europeans, European colonial governments, and flora and fauna. She used her own material for lectures, articles, and more substantial publications that she wrote either jointly or solo. We draw accounts of river transport from her book *Travels in West Africa, Congo Français, Corisco and Cameroons*, published in 1897 in London.[19]

Most notably, in June 1895 Kingsley sailed up the Ogooué River, Gabon's main waterway, on the sternwheeler steamboat *Éclaireur*. She departed from Cape Lopez on the coast, passed Lambaréné (a town that from 1913 would become the site of the hospital founded by future Nobel Peace Prize laureate Albert Schweitzer) and proceeded to the French mission at Talagouga, near the town Ndjolé. The *Éclaireur* belonged to the shipping company Chargeurs Réunis, subsidized by the French government. After securing the necessary permits to proceed into the heart of the colony, Kingsley set out. On the first few days of the trip—from June 5 to June 8—she traveled on the boat *Mové* and then continued on the *Éclaireur* to the end of the voyage on June 25.[20]

The journey to Talagouga was fairly smooth. Along the way, Kingsley was able to study the other passengers and crew, but above all, she was able to admire various aspects of the Gabonese landscape, observing with a scientific eye the flora parading by her on the banks, but also responding emotionally to the sights along the way. For example, she described the Ogooué

River, which flows in a deep cleft between perpendicular walls not long before Talagouga, as 'a gloomy ravine'.[21]

Familiarized with the environment of the French mission, Kingsley wanted to ascend the Ogooué rapids past Talagouga and Ndjolé to Kondo Kondo Island. The ascent after Ndjolé could only be done by canoe. She said she was willing to rent a crewed canoe for 100 francs, paying for working expenses, food, wages, and other costs as well. Monsieur Gacon, a member of the Mission Évangélique in Talagouga, provided her with a canoe in good condition and two English-speaking Igalwa guides, M'bo and Pierre. The former had accompanied a couple of years earlier the two missionaries Elie Allégret and Urbain Teisseres of the Société des Missions Évangéliques de Paris from Franceville in Gabon to Brazzaville, the current capital of the Republic of the Congo, and then down the Congo River. Gacon proposed to complete the crew with six Fang men, but they refused to go up the river past Ndjolé because of tribal issues, 'because they were certain they would be killed and eaten by the up-river Fans'. Two more Igalwa were then hired.[22]

Kingsley also provided a rather detailed description of the vessel, crew positioning, and space organization:

> On we paddled, M'bo the head man standing in the bows of the canoe in front of me, to steer, then I, then the baggage, then the able-bodied seamen, including the cook also standing and paddling; and at the other extremity of the canoe … stands Pierre, the first officer, also steering; the paddles used are all of the long-handled, leaf-shape Igalwa type.[23]

Thus, note the special care with which the best means of transportation is chosen, given the difficulties involved. And by 'means of transportation', we do not mean merely the object 'canoe' and its sound condition and the rational management of space but also the preparation of the crew. Specifically, head man M'bo and first officer Pierre were proposed by Gacon based on skill sets that were appropriate for Kingsley's itinerary. Both spoke English, plus M'bo already had experience accompanying European travelers in the wilderness and on African rivers. Therefore, the transportation in this case was arranged with all the trimmings.

Beyond the town of Ndjolé, the Ogooué River was squeezed between steep rock walls, and its waters ran fast with dangerous rapids. Kingsley and his crew encountered unforeseen obstacles to navigation. The specialization of M'bo, Pierre, and the crew was therefore indispensable in overcoming these difficulties. There were whole trees that had fallen into the river, emergent rocks, and whirlpools. These were all obstacles that required experience to negotiate, especially when the group had to move at night. The large trees that had fallen into the river could not be distinguished in the darkness, so the keen ears of the guides were fundamental in distinguishing the sound of the current flowing among the branches. And dodging whirlpools required considerable skill in the use of paddles and poles, as well as considerable swiftness and agility in moving into the proper position on the canoe to handle a sudden challenge. Kingsley's crew proved equal to the task.[24]

As in the case of Brocchi in a felucca on the Nile, Kingsley's adventurous river voyage allowed her to appreciate the natural setting firsthand, experiencing a true form of experiential tourism *ante litteram*. First, being on a canoe in the heart of Gabon allowed her to behold breathtaking natural spectacles, which she was able to describe to her readers effectively in *Travels in West Africa*. For example, the large dark rocks that emerged from the rushing current of the Ogooué and were struck by sunlight had 'a soft light blue haze round them, like a halo'. Kingsley confessed this sight, complemented by the forest on either side and the snow-white beaches, 'was one of the most perfect things I have ever seen'. In similar tones, she described the twilight over Gabon, the sky tinged with 'that divine deep purple velvet which no one has dared to paint'. After the twilight, the waters of the Ogooué and the canoe found themselves plunged into a darkness illuminated by 'great stars blazing high above us' and 'be-gemmed with fire-flies'. These visions were crowned by the beauty of the night when Kingsley and her crew camped on Kondo Kondo Island:

> The sun went down and the afterglow flashed across the sky in crimson, purple, and gold, leaving it a deep violet-purple, with the great stars hanging in it like moons, until the moon herself arose, lighting the sky long before she sent her beams down on us in this valley…. Around, on all sides flickered the fire-flies, who had come to see if our fire was not a big relation of their own, and they were the sole representatives, with ourselves, of animal life.[25]

Moreover, some particularly difficult and dangerous stretches of the ascent of the Ogooué amounted to a bona fide white-water rafting experience, similar to that described by Acerbi on the Tornio. Kingsley described the crew's earnest struggles on one particularly difficult occasion 'and many a wild waltz we danced that night with the waters of the River Ogowé'. We report a particularly pathos-filled passage from that night's adventure:

> About 9:30 [p.m.] we got into a savage rapid. We fought it inch by inch. The canoe jammed itself on some barely sunken rocks in it. We shoved her off over them. She tilted over and chucked us out. The rocks round being just awash, we survived and got her straight again, and got into her and drove her unmercifully; she struck again and bucked like a broncho [*sic*], and we fell in heaps upon each other, but stayed inside that time….

We sorted ourselves out hastily and sent her at it again. Smash went a sorely tried pole and a paddle. Round and round we spun in an exultant whirlpool, which, in a light-hearted, maliciously joking way, hurled us tail first out of it into the current.[26]

Note also that Kingsley's description of actions made is always in the first-person plural. She certainly did not have the same expertise or physical strength as M'bo, Pierre, or the rest of the crew who handled paddles and poles, but we can conclude with certainty that she was a first-person participant in the maneuvers, refusing to be a deadweight for the canoe.

There are thus all the traits to define the ascent of the Ogooué as a form of experiential tourism for Kingsley: navigation with the goal of seeing something new, in this case the area above Talagouga and Ndjolé and Kondo Kondo Island; active participation in local navigation techniques on an indigenous vehicle and with an indigenous crew; and immersion in the natural scenery, moreover with stops in local villages. It is no coincidence that nowadays some companies offer various nature tours of Gabon with local guides along the Ogooué.[27]

In three different geographic settings and eras, however, the river transportation experiences of Acerbi, Brocchi, and Kingsley are united by their strong experiential character, in which local means of transportation, mobilized under conditions of near urgency, afforded the three travelers experiences of nature and contact with local people that would not otherwise have been possible. In this, De Bruijan's 'cruise' on the Volga in the early eighteenth century differs sharply from the other three case studies, aligning more closely with leisure tourism, where nature, largely domesticated, was mainly a backdrop.

One characteristic shared by the transportation systems in all case studies is the necessary involvement of specialized personnel to operate the craft employed in each specific case. On the journeys of Acerbi and Kingsley, this personnel was particularly critical in negotiating natural obstacles, adding the thrill of battling white water to the experience of the journey.

Let us now consider some cases of sea transportation over both long and short distances. De Bruijn gives us an initial idea of the time needed and the difficulties encountered in the early eighteenth century in covering medium or long stretches by sea. In the summer of 1701, he left from the Hague, stayed in Amsterdam for two days, and then embarked on a ship from the island of Texel for the first leg of the voyage to the Russian port of Archangel on the White Sea, circumnavigating the Scandinavian peninsula off Norway. The first leg lasted from August 1 to September 3, with some difficulties near this initial destination, whence he needed authorization to proceed.[28]

The long duration of this voyage was not only a question of the relatively slow speeds of early eighteenth-century ships. There was also no organized international transport network dedicated to those traveling for pleasure (or other reasons). From Amsterdam to Texel, De Bruijn traveled by 'ordinary conveyance'. For the trip from Texel to Archangel, he embarked on the flyboat *John Baptist* with a crew of eighteen men. The flyboat was a light vessel used mainly for mercantile purposes. The *John Baptist* on which De Bruijn embarked had no declared military missions but mounted eight cannons in case of need.[29]

International tensions were keenly felt at Dvina Bay in the White Sea, before sailing up the delta to Archangel. On the morning of August 31, 1701, De Bruijn counted twenty-one foreign ships moored near the mouth of the Russian River: eleven Dutch, eight English, and two Hamburgers. Despite the favorable weather, all the ships waited restlessly for the arrival of specialized pilots because the Russians had removed all sea markers to thwart the Swedes, whose ships had arrived a few weeks earlier 'and alarmed all the neighborhood'. The Great Northern War had broken out the previous year between Russia and the Swedish Empire and would last until 1721. The long conflict would at various times involve numerous European states as allies of one side or the other and would ultimately stem Sweden's expansionist ambitions. Without guidance from local captains familiar with the delta, a number of ships attempting to sail up the Dvina between August 31 and September 1 damaged their hulls on the rocks or ran aground.[30]

The climate of war also obliged De Bruijn and all men entering Russia to stop and wait at Novodvinskaya Fortress for authorization to enter the country. Tsar Peter the Great had begun construction of the fortress just as hostilities with Sweden began. De Bruijn wrote 'Here also they were getting ready three branders, and a chain of ninety fathoms ... to obstruct the Swedes, who had been every day dreaded, since their last expedition'. Only when the commanding officer of the fortress gave them permission to proceed—and 'entertained us with a glass of brandy'—were they allowed to continue to Archangel.[31]

The details of De Bruijn's voyage from the Netherlands to Russia do not make particular reference to the natural scenery, but they do convey the difficulties in transportation and access to the various countries that were encountered during this long international voyage. His very comfortable trip on the Volga was in a much quieter and more organized setting: a river route touching population centers and important Russian ports, organized with some care. But this was not the typical experience of international travelers who ventured into increasingly wild natural environments, where the uncertainties and dangers increased exponentially. De Bruijn was aware of this. Not surprisingly, he reported in the opening of his book that relatives and friends had tried to dissuade him from leaving for Asia and the East Indies, pointing out 'the consequences and inconveniences of such a project'.[32]

In De Bruijn's account of his voyage to Archangel, we noted the need for specialized figures in navigating certain areas, especially when it came to guiding foreign travelers. In this case, they were the pilots who guided ships through the Dvina delta after the sea markers had been strategically removed to hamper a possible Swedish invasion. Specializations thus existed for short or long sea routes analogous to their counterparts in river transport on wild or nearly wild stretches.

Almost a century later, in August 1796, the ship *Duff* sailed from London under the command of Captain James Wilson. She was bound for the Pacific islands with a group of evangelical missionaries from the London Missionary Society who had hired Wilson specifically. The voyage had a completely different purpose than De Bruijn's, not having as its ultimate goal artistic research or a desire for new experiences as in the case of the Dutch painter. In addition, the crew and passengers of the *Duff* had not the opportunity to either enjoy luxuries or focus with awe on many particular beauties of nature. Nonetheless, a description of the *Duff*'s adventures, the lands visited, and the peoples encountered was printed in London in 1799, bringing together as documentary material the journals of the ship's officers and the missionaries themselves. It includes interesting and faithfully described not only accounts of the customs and nature of the Latin American coast and especially the Pacific islands but also valuable details on the management of the voyage. For this reason, analysis of some aspects of the *Duff*'s voyage may be useful to better understand the difficulties of sea travel.[33]

The *Duff* carried thirty-nine missionaries: thirty men, six women, and three children. The ages of the adults ranged from twenty to thirty-seven, with the exception of a sixty-four-year-old woman who was the wife of one of the missionaries. All three juveniles were children of missionaries on the *Duff*. One was twelve years old and was the son of an ordained minister. The others were brothers aged two years and sixteen weeks, respectively, and were the children of an Indian weaver. From the many applications for the mission, the society had selected those most suitable both for moral and religious conduct and the usefulness of their profession. Thus, in addition to four ministers, there were carpenters, tailors, weavers, shoemakers, a surgeon, and others. Onboard with the thirty-nine missionaries were Captain Wilson, his nephew William (first mate), and twenty-one other men, including second and third officers, cooks, and sailors. Of the crew 'about half were communicants'. Based on these numbers the initial provisions were organized as follows: 'Tea, one pound per month each mess for the men, one pound and a quarter for the women. Sugar, two pounds per week each mess. Butter, one pound and a half per week each mess. Cheese, four pounds per week each mess'.[34]

While sailing, the *Duff* had to cope with obstacles of various kinds. For example, on the evening of August 15, 1796, within days of leaving the Thames, not far from Beachy Head along the southern coast of Britain 'one of His Majesty's sloops of war hailed and informed us that a French lugger was cruising somewhere near; bid us keep a good look-out, and hoist a light if we discovered her, whilst they ran in nearer the shore'. Keep in mind that in 1796 France—a Republic then since only September 1792—was experimenting with Directoire rule. France was relatively stable on the home front (at least until the coup in November 1799), especially in comparison with the previous period of the 'Reign of Terror'. On the foreign front, however, the waters were far from calm: most European monarchies, including Britain, had organized themselves into the first anti-French coalition, which would last until 1797. Although still close to home, we may presume that the *Duff* was already at risk because of the war that pitted the two countries against each other.[35]

In February 1797, the discomfort was psychological in nature. The *Duff* was in the middle of the Pacific Ocean, and on the 21st, it had been sailing for ninety-seven days since their last landfall (Rio de Janeiro). During that journey (the ship logged 13,820 miles), the *Duff* had encountered only one other vessel, about a week out from Rio. Otherwise, they had not encountered any other ship, had not reprovisioned anywhere, and, keeping away from coasts and islands, had not seen land again. It is true that there had so far been no mishaps on this leg—'the Lord, in providence, has supplied our necessities in a most wonderful manner'—yet the uninterrupted sight of 'a vacant horizon, and the familiar objects the sea daily presented to our view' were wearisome. Indeed, when they finally sighted Tubuai Island, the largest in the Austral Islands 'discovered' by James Cook twenty years earlier, there was great jubilation.[36] Sailors and other types of passengers spent long periods of time away from their homelands and families. Historiographers have shown significant interest in the many aspects of these experiences, from food, hygiene, and working conditions to behavior and sexuality.[37]

At other times, the sea and the weather were not quite so clement. The previous year, in the second half of October 1796, still in the Atlantic Ocean before arriving in Rio, the *Duff* had encountered bad weather. For example, on day 21 'a heavy squall came on when the missionaries were engaged between decks in evening prayer; and the ship heeling on a sudden, the lee scuttle being open, the water rushed in like a torrent'. On day 27 'the weather became now very unsettled, the winds variable and squally, attended with much thunder and lightning'. In early April 1797, they sighted Palmerston Island, a coral atoll in the Pacific Ocean, and decided to try to go ashore to stock up on coconuts. They approached within two leagues of the shore, put the pinnace and jolly boat in the water, but could not get ashore, at first 'on account of the surf breaking high on every part of the surrounding reef' and

later because of 'a squall of wind and heavy rain coming on'. Equally unsuc-
cessful was the second attempt, although the weather had improved. Eventually, the second mate, who was Tahitian, and one of the sailors ventured
into the water partly swimming and partly walking on the reef, cutting themselves in various parts of their bodies but managing to make it to shore.
Later, the boats found a passage deep enough in the reef and managed to
reach the beach.[38]

If the *Duff* went through difficult experiences animated by a missionary
spirit, the *Beagle* did so for scientific reasons. We have other interesting
accounts of sea navigation from Charles Darwin's notes during the famous
voyage from late 1831 to late 1836, which we have already had a chance to
analyze from other perspectives. The *Beagle* encountered numerous difficulties in navigation, which Captain Robert FitzRoy had to promptly remedy.

For example, when the *Beagle* easily reached Cape Horn a few days before
Christmas 1832 with the help of a 'fine easterly breeze', it found a very different situation once it had rounded it. Famous is the description of how
Cape Horn 'demanded its tribute' from the *Beagle* with a gale that Darwin
suggestively described: 'Great black clouds were rolling across the heavens,
and squalls of rain, with hail, swept by us with extreme violence'. The storm
forced FitzRoy to seek shelter on Christmas Eve itself in a small cove in the
area, where miraculously the *Beagle* found itself in calm waters protected by
the mountains.[39]

Nowadays, tour operators offer visits to natural sites in Tierra del Fuego
and cruises off Cape Horn billed as following in the footsteps of Darwin and
FitzRoy. For example, a trip 'from Ushuaia to Punta Arenas on a small-ship
expedition cruise' lasting five days can be had for upward of US$2,000. Of
course, there is the substantial difference that these tours are organized,
combining contact with nature and local history with the safety of modern
instrumentation and transportation.[40]

Of great interest for this part of our study is Darwin's description of the
Beagle Channel when he had a chance to cross it in January 1833. The Beagle
Channel is about 240 kilometers long and crosses the Tierra del Fuego
Archipelago, cutting across the southern part of present-day Argentina and
Chile. It derives its name from the HMS *Beagle*, which reached it under Fitz-
Roy's command during the hydrographic survey of 1826–1830, a voyage that
preceded the one Darwin took part in.[41]

In January 1833, the *Beagle* dropped anchor in the sea between Navarrino
Island and Lennox Island, and the captain decided to sail into the channel
to leave off some natives picked up on the *Beagle* during the voyage at a
certain point and then continue the exploration. Darwin was part of the
group of twenty-eight people aboard three whaleboats and a yawl under
FitzRoy's command who sailed into the eastern mouth of the channel on the
afternoon of the 19th. During the twenty-day excursion, Darwin and his

companions immersed themselves in the channel's majestic landscape: they saw whales and icebergs, admired the forested slopes below towering snow-capped peaks on both sides, and observed several glaciers descending to the shore. They made numerous stops, camping in the inlets along the channel, deeply immersed in the local environment. They also came into contact with various indigenous communities, not always on peaceful terms.[42]

Darwin thus had an intensive experience of the southernmost reaches of South America. During one of his night watches, he penned this in his journal:

> There is something very solemn in these scenes. At no time does the consciousness in what a remote corner of the world you are then buried, come so strongly before your mind. Every thing tends to this effect; the stillness of the night is interrupted only by the heavy breathing of the seamen beneath the tents, and sometimes by the cry of a night bird. The occasional barking also of a dog, heard in the distance, reminds one that it is the land of the savage.[43]

Darwin's experience was a precursor to the interest shown by tourists today in the Beagle Channel, with boat trips varying by type of boat, stops, duration, and price. They allow travelers to connect with the natural scenery, wildlife, local history, and indigenous culture through a wide variety of nature-based experiences such as penguin watching and the promise of retracing Charles Darwin's footsteps.[44]

Although these tours do not necessarily correspond to Darwin's route through the Channel, we nevertheless find a strong similarity with the river tours we have examined in Egypt, Gabon, and Northern Europe, where good organization tended to minimize possible discomfort but, above all, the opportunity to connect with the natural environment and local traditions.

Marianne North, a famous English painter of botanical subjects whom we introduced in Chapter 1, also offers us information on water transportation. In her case, however, we consider sea transport over short distances. During her stay in the Seychelles between October 1883 and early 1884, which we had the opportunity to describe as regards her perception of nature, North used several boats to travel from island to island and conduct her own research on the local flora that she would paint. However, there did not seem to be regular transportation service from one island to another, whereas land transportation, as rudimentary as it was, had some semblance of a fare structure, as we shall see later in this chapter. Thus, North, who was an exceptional guest anyway, always used boats provided by local authorities or notables.

For example, on October 22, 1883, North traveled from the main island of Mahé to the island of Praslin. Commissioner Barkley, who resided on Mahé, put his own small sailing boat at her disposal. On a windless evening, it took her three or four hours to 'slowly' travel twenty-five miles of sea. Later, just off the coast of Praslin, North transferred into the 'little boat' belonging to her hosts on that island, the family of Doctor Hoad, a local physician, who brought her ashore. As we have seen in previous examples, a trip arranged with the means at hand and under less-than-optimal conditions, here the lack of wind, turned out to be an excellent experience of contact with the natural landscape and a genuine approach to local life. North landed on Praslin at midnight 'in the lovely moonlight', and as she 'jumped on the sand' two friendly dogs greeted her by licking her hands and face, almost knocking her over, to her delight. The Hoads had sent a kind boy of thirteen to welcome her and guide her to their new abode, which was not a 100 yards from the shore, shaded by tall mangos and other trees. The sea was visible through the fronds, the blue waters taking on a sapphire hue at midday. North, generally anything but lavish with praise, characterized her stay in Praslin as 'a real good time'.[45]

Moving frequently from island to island in the boat used by Hoad for visits, North got to experience different transportation situations set up at the time, but in all of them, she was able to find interesting aspects that allowed her to learn more about the nature and people of the Seychelles.

For example, from Praslin, she visited Curieuse Island, which was a half-hour paddle trip. A poor old Englishman lived on that island, which was private; he had married the heir to the deed. Approaching from the shore, North could admire the 'long double avenue of large lilies and crinums' that bordered the road to the Englishman's old house, which was 'picturesquely dilapidated'. Here, too, man-made nature harmonized with the natural beauty of the setting, which North had a chance to visit during her stopover.[46]

North also went to Aride Island, which had been purchased for 120 pounds by a wealthy Creole from Praslin. As its French name suggests, the island is rather barren, as well as rocky and rounded, but the owner 'kept people there' to grow banana and other tropical fruit trees on what little flat ground there was. As the painter noted, the owner of Aride Island also had some income from hunting local terns and selling them to the people of Mahé for culinary purposes, an activity that would seriously damage the island's ecosystem in the decades following North's visit. The landing did not look easy, but the owner sent for a pirogue to take North and the Hoads—and Toby, Mrs. Hoad's little dog, firmly in his mistress's arms—'over the wonderful opal sea'. The travelers transshipped from Hoad's boat and waited on the pirogue until they were driven to Aride Island 'with a rush, on the top of the wave'.[47]

A third example. North accompanied the lone Doctor Hoad on a medical tour of La Digue as well. She described it as 'a large fertile island containing six hundred people', where coconut palm cultivation and oil production and export flourished, as evidenced by the numerous boats at anchor. (Actually, upon closer investigation, North found that water and food supplies were scarce.) The island was reached via a very narrow and dangerous passage through the reef, which was difficult to negotiate in fair weather and impossible in bad weather. North knew that one of Hoad's predecessors had drowned in that passage. But once again, despite the difficulty of transit by sea, the painter knew how to find the silver lining and was enraptured by the wonderful colors of the corals through the crystalline waters.[48]

We have a description of the crew of Hoad's boat, in which the doctor traveled from island to island for his visits and offered rides to North. There were four 'capital sailors' who rowed. The young head man was named Emile, born in Madagascar, for whom North clearly had a soft spot, despite his moody and turbulent character: on more than one passage she paused to admire his 'fine forehead and straight nose', his 'grand profile', his youth and energy. There were also two Seychelles natives and one 'pure African'. The men were strong, jovial, and chatty. North was happy with them and admired their 'amphibious' nature as natural navigators having no fear of the sea.[49]

We make mention of the four sailors in this Section 3.2, and not in Section 3.4 devoted to the relationship between travelers and local guides, because the description of Emile's character confirms the fortuitous nature with which transports between the various Seychelles islands were organized and how, beyond reliance on the undoubted expertise of the locals, there was no itemized contract detailing the crew's responsibilities to the 'client' and any guests. In fact, Hoad told North about a significant episode that had involved him directly:

> Once he [Emile] had imprudently tried to pass through a dangerous passage among the rocks and surf, and the boat was turned right over. The doctor found himself swimming in pitch darkness among the sharks, and heard Emile fighting with the others as to whose fault it was, never thinking a moment of helping the lost doctor, who fortunately was able to swim half a mile to the shore.[50]

Having analyzed water transportation, let us now consider land transportation. Here, too, we notice remarkable differences depending on the era and the territory. Let us take as our first example the young Linnaeus's trip to regions in the Kingdom of Sweden in 1732: his *Lachesis Lapponica* reveals the difficulties he encountered in using a tired horse unsuited to the rough trails through the Scandinavian wilderness. For example, toward the end of May, he was in the city of Umeå and decided to visit the Ume River Valley (Lycksele Lappmark) on his own. He arrived there following a route with no posthouses. He thus did not have the opportunity to obtain a fresh horse, suitable for the road ahead, 'which is no small convenience to a stranger traveling in Sweden'. Moreover, Linnaeus's horse had a poorly padded saddle with a rope instead of a proper bridle.[51]

Not surprisingly, Linnaeus soon encountered a certain incompatibility between his mount and the increasingly narrow road that was so uneven 'that my horse went stumbling along, at almost every step', proceeding with exasperating slowness and putting the rider's life at risk several times. There were virtually no humans in his vicinity until he reached a village in which he stayed overnight but again was unable to change horses. When he resumed the journey the next day, the situation was worse than ever and the road 'was indeed the worst I ever saw, consisting of stones piled on stones, among large entangled roots of trees'. Crevices and deep potholes had filled with water from the recent rain, birch branches hung soggy and weighed down along the route, mighty pine trees had been felled by the fury of the storm and now lay on the road, making it arduous to proceed. The Ume River Valley was innervated with deep streams and creeks. The locals had built bridges over them but they were in such a ruinous condition 'that it was at the peril of one's neck to pass them on a stumbling horse'.[52]

Although the forest unfolding around him awakened his attention as a naturalist, Linnaeus did not take his wilderness experience in the Ume River Valley very sportingly, unlike Acerbi's boat adventures on the Tornio River or Brocchi's and Kinglsey's in Africa in the following century. Linnaeus complained quite a bit about this stretch, also cursing the people's neglect of the region's roads and bridges. He reported that there was a hamlet whose church was seven miles away over an 'execrable' road, obliging churchgoers to set out on Friday morning if they wanted to be in church on Sunday. Needless to say, the village was not well represented during services. Exaggerations aside, the importance of an overland travel network with well-maintained roads, solid bridges, properly spaced and located posthouses, good horses, and other services was especially critical when crossing territories with little human presence.[53]

Very interesting in this regard are the letters written by the brothers Paolo Greppi (1748–1800) and Giacomo Greppi (1746–1820) to their father, Antonio (1722–1799), count, important banker, merchant, and diplomat in the service of Habsburg Lombardy. The two brothers took a tour of Western Europe from 1777 to early 1781 both for professional growth and to strengthen trade relations with European courts and mercantile societies, at times traveling separately but often together. The letters to their father contain a great deal of information on the difficult traveling conditions through the countryside between one town and the next, with poor roads and signage and total vulnerability to weather (dust and sun in summer; rain, mud, snow, and ice in fall and winter). The carriages could only make slow headway and were at risk of accidents and damage, with frequent need for repairs further slowing progress. The letters indicate that the road network tended to deteriorate further as they proceeded southward. Other questions regarded the recruitment of capable servants on each stretch, the choice of lodging, and the risk of highway robbery in many of the areas they traversed.[54]

The importance of an adequate transportation network also emerged at the beginning of the next century in the notebooks of Carlo Amoretti in his descriptions of the Italian peninsula. Until the arrival of French troops in Milan, Amoretti had been the secretary of the Patriotic Society, founded by Maria Theresa, a body dedicated to boosting agriculture, animal husbandry and manufacturing in Austrian Lombardy. In 1808, during the brief Napoleonic Kingdom of Italy—from 1805 to 1814—he was appointed a member of the Council of Mines. He was thus a man who worked in close contact with the rural and natural world, as already revealed in Chapter 2.[55]

Descriptions of the transportation network and its relation to the natural environment emerge in the notes Amoretti took during a trip he made from Milan to Naples and back between early November 1801 and early May 1802. Amoretti stopped for some time in Naples and in Rome during the trip. In spite of these social and cultural interludes, we may infer that the main objective of his trip was to gather information about Italy's natural resources. Indeed, his keen eye was always focused on the composition of the natural setting, and even when he was in the city, he did not lose sight of his scientific interests. For example, when he was in Naples, he made a point of climbing Vesuvius to observe the volcanic crater.[56]

In November, he traveled through the countryside of Tuscany and Latium. Some areas were rich in cereal crops, vineyards, olive groves, and other crops and dotted with cattle, sheep, pigs, farmhouses, and villages, while others were wilder with woods, ravines, and uncultivated land. In his notes on the road between Florence and the town of Poggibonsi, Amoretti recorded solid bridges to facilitate stream crossings and roads finished in paving stones on uphill stretches, 'without which the rains would furrow them'.[57]

Amoretti traveled together with a servant and other passengers in carriages drawn by mules. Mules move much slower than horses, but they are hardier and better suited to rough terrain. However, although the road network in central Italy was generally well maintained and equipped with horse-changing stations, it did pose a number of challenges. It was frequently rough and uneven, and travelers occasionally encountered particularly steep hills, forcing them to get out of the carriage and proceed on foot, with the coachmen and lightened carriage following behind. In at least two stretches, one in Val d'Orcia after Siena and the other a few days later outside Viterbo, Amoretti and his traveling companions had to walk several miles before being able to get back into the carriage. In another case, the road was so steep that oxen had to be found to help the mules. Amoretti generally took advantage of these inconveniences to study his surroundings more calmly and to make small side trips to points that interested him.[58]

The importance of suitable mounts was also quite evident in Brocchi's explorations in the Egyptian desert during the 1830s. This was quite a different context from Amoretti's more comfortable central Italian campaign and from Linnaeus's experience in Lapland. Brocchi traveled by felucca on the navigable portions of the Nile, but his geological surveys for the *Wali* often focused on inland locations. The ideal way to reach them was obviously by dromedaries organized in caravans.

For example, after landing in Qena, he had to travel overland to reach the areas near the Red Sea where he was to conduct his geological and mineralogical surveys. Here the dromedary was essential. In addition to their incredible resistance to thirst, he noted: 'A good dromedary can go a hundred miles in 24 hours. At the caravan pace it makes two and a half an hour; at a trot, four'. He observed, however, that dromedaries were not unaffected by heat, despite their considerable endurance, and that with the coolness of the night, they went faster. Brocchi was fascinated by this animal and sought more information from the locals about its origins, cross-referencing it with the scientific knowledge he had gained in his studies. In truth, there were still some gaps in that knowledge because he seemed to believe that a dromedary was just a young Bactrian camel and would grow another hump as it matured, but he was very clear about its importance as a means of transportation and often calculated distances as a function of the time it would take one of these animals to make the trip.[59]

In other case studies, the terrain was too rugged even for the use of mounts. For example, this might be the case where there are no roads or paths and the vegetation is very thick. A good example is the Scottish botanist Anna Maria Walker, whose ascent of Adam's Peak (Sri Pada) in Sri Lanka in early 1833 was described in Chapter 1. While she went on foot where she could, she was often transported in a 'Kandyan moonshull, something resembling a hammock, swung on a pole', a means of transportation used by Sinhalese notables. Compared to the Western-style mounts that were increasingly popular among the natives, the 'moonshull' had the advantage of being usable on the roughest roads because porters could pass over any terrain.[60]

With a certain amount of humor, she also mentioned some of the inconveniences caused by the clash of her British pragmatism with the beliefs of the locals:

Our route the whole way was a precipitous ascent up the bed of a torrent at present quite dry. In many places, my position in my little palankeen became exceedingly awkward and uncomfortable, my feet being higher than my head. I tried to persuade the bearers to turn the vehicle and carry me backward; but this they considered unlucky, and could not be prevailed on to do.[61]

Marianne North found herself in a somewhat similar situation during her time in the Seychelles in late 1883 and early 1884. Newly arrived on the island of Mahé, she was told that it was impossible to walk outside of town, climb the slopes on foot, and make her way through the dense vegetation. She let herself be persuaded for her first two expeditions to travel 'with four bearers and a chair' at a cost of 15 shillings per day. Soon, however, he realized that he could cope with the local topography on foot without any particular problems, the real obstacle being the men placed at her disposal. She thus stopped using any land transportation other than her legs. She was of the mind that only on the first expedition had the sedan chair been of any benefit. She had gone with a local scholar to see a deformed coconut palm specimen, and the route had involved a particularly rugged stretch of the island's coastline, 'up and down from one sandy bay to another'.[62]

North, of course, was familiar with more technological means of transportation such as the train. When she visited New Zealand in early 1881 after a long stay in Australia, she made use of local trains. While not overly enthusiastic, she did recognize the fact that they made it possible to reach places in one hour that fifty years earlier had required three days of travel over difficult terrain. When she visited South Africa between 1882 and 1883, she made several trips by rail, alternating with post carts and carriages. By the mid-1880s, Western culture was firmly entrenched in both New Zealand and South Africa, bringing larger urban centers and an expanded transportation network. These were hardly the Scandinavian forests through which Linnaeus and his horse slogged 150 years earlier.[63]

Train travel did not preclude enjoyment of the natural setting. North beheld breathtaking landscapes through the window, where domesticated nature gave way to wilderness. Commenting on her approach to the town of Invercargill on the south coast of the South Island of New Zealand, she wrote that 'the views passing through the low boggy country were most picturesque, with rich browns and purples, streaks of water, and distant mountains, long-legged blue birds standing on pedestals of stakes in the foreground'.[64]

During the years of his travels, Italian priest Luigi Marchelli was able to cover long distances in a short time thanks to the train and the rail network that was developing at that time. In addition to visiting cities that until a few years earlier had belonged to different countries, he was also able to experience a revolution in travel times. We find three aspects in this regard in his notebooks: the first regarded the emotional experience of the passing landscapes and his impressions of his traveling companions; the second, of a more historical and social nature, regarded travel times and the development of the infrastructure on new routes, some which had just been opened; and the third, more along the lines of economic history, regarded the cost of train travel. There was thus a synergetic relation between the railroads and tourism: in some cases, it was the infrastructure to stimulate tourism, especially in cities that already had rail connections; in other cases, it was tourism that stimulated the development of the railways to connect seaside, thermal, or mountain resorts.[65]

3.3 Food and Lodging

Accommodations are among the fundamental elements of the modern tourism industry and play a particularly important role as experiential elements in tourism modes such as nature tourism, rural tourism, and pescatourism, among others.

An aspect that has been well investigated by scholars of tourism history relates to hospitality facilities and their evolution over the long term. Professional hospitality has developed over time both in relation to the needs of those traveling for work (trade, scientific expeditions, etc.) and regarding tourism, with an ever-increasing demand for lodging also in the environments we have been studying. The type of accommodation is also an essential variable in the type of tourist experience one seeks. For example, sleeping in treehouses, tents, or in local villages during experiential tourism adventures in forests, savannas, deserts, or other ecosystems appeals to the traveler's desire to have contact with novel natural contexts and the lifestyles of the local inhabitants.[66] At the same time, the 'wild' element sought today in most cases is controlled by tour operators or tourism professionals. This sort of tourism takes place in settings that are different from many, although not all, of the travel experiences in the eighteenth and nineteenth centuries that we have studied.

How did the eighteenth- and nineteenth-century travelers who sought contact with nature in their experiences fare in this regard, not knowing that they were ahead of their time? As we did with transportation, we now examine this aspect of tourism, with relatively long time frames and lower levels of safety and hygiene than we are accustomed to today. Some of the travelers who provided information on transportation modes in Section 3.2 also share a great deal with us as to how they ate and slept.

We may again turn to young Linnaeus in Lapland in 1732. During his arduous trek through the Ume River Valley described previously, he came to a village where women were intent on some work and he asked them what he could eat for supper. They served him some grouse breast 'shot and dressed' the previous year. The appearance was not inviting, and Linnaeus expected no better of the taste. However, he was pleasantly surprised and described it as 'delicious'. With some rhetoric, but rightly so, he commented:

I wondered at the ignorance of those who, having more fowls than they know how to dispose of, suffer many of them to be spoiled, as often happens in Stockholm. I found with pleasure that these poor Laplanders know better than some of their more opulent neighbors, how to employ the good things which God has bestowed upon them.[67]

Referring to natural resources of plant origin, and edible berries first and foremost, in recent times Matti La Mela has analyzed the concept of *allemansrätt*, 'a Nordic tradition of public access to nature' in Sweden and Finland, focusing especially on the regulatory issues related to property rights in the second half of the nineteenth century. Already in the case of Linnaeus we can sense the importance of natural resources and free access to them in sparsely urbanized areas of Scandinavia, a prelude for the framework analyzed in depth by La Mela.[68]

Linnaeus also described the quite elaborate preparation of the grouse. The meat was cut into pieces and cleaned, then salted, laid on a bed of flour, and placed in an oven to be dried gradually. Once the process was finished, the meat was hung from the rafters in the houses and could keep for up to three years before being eaten. He stayed in the village until the next noon because of rain and thus also provided a description of the overnight stay. He slept on a bed with pillows stuffed with reindeer hair; under the sheets was the skin of a reindeer with the hair side up as a soft mattress cover. Linnaeus did not complain.[69]

Carlo Amoretti's notebooks of seventy years later are also quite interesting in this regard, when he traveled the countryside of Tuscany and Latium in November 1801, moving from wilder to more typically agricultural scenery. If Linnaeus had been stranded in an unfamiliar Lappic village because of rain but, on balance, had found some comfort there, Amoretti could not say he was as fortunate when he arrived in Poggibonsi, a town not far from either Siena or Florence, 'through the fog and rain' with his traveling companions. The inn in which he stayed was 'large, but of deplorable quality, poorly structured, and dirty'. The travelers were housed in two 'horrible rooms' and Amoretti, in order to sleep more comfortably, was forced to join two mattresses on which he placed his 'sack bed'. To make matters worse, his fellow guests at the inn did not inspire Amoretti with great confidence, so he and his fellow travelers kept away from them. They fared better with the food: in addition to seasonal vegetables such as cauliflower and the provisions that Amoretti and companions had with them (not indicated in detail), the inn seemed to be a distribution point for fish coming from the nearby Maremma, a coastal area, and then headed to the various inland towns. They were thus able to eat good fish. Also, as Amoretti noted, Poggibonsi was an area of vineyards and wines, and we may infer that they were served as well.[70]

The next evening, Amoretti found a roughly similar situation at the inn in Buonconvento (now rated after a meticulous evaluation process as one of the most beautiful villages in Italy by the Borghi più belli d'Italia association, established in 2001 by the Consulta del Turismo dell'Associazione Nazionale dei Comuni Italiani [Tourism Council of the National Association of Italian Municipalities]). In that distant 1801, Amoretti made no particular mention of the beauty of the village, commenting only that the surrounding countryside, planted with wheat and flax fields, vineyards, and olive groves, was not worthy of any particular praise. The inn was not bad, and the group was provided with two rooms, although they were not very large and could be accessed only by passing through the rooms of some of the sick. For food, they again found fish and local produce, such as cauliflower and eggs.[71]

In Radicofani, however, another ancient village in the Sienese hills, Amoretti found a more remarkable lodging. Upon seeing the building in which he would stay overnight, Amoretti was favorably impressed:

> I, who had found very few inns in populous villages, thought I would not find shelter on the summit of some deserted mountain; and I was well surprised when I found myself in a magnificent palace, and in two good rooms, with furniture and utensils, and very proper linen; so that I had no need to avail myself of the sleeping bag. I then learned that it was a Medici palace built with all sumptuousness, because they used to sojourn there on the occasion of their hunts.[72]

This was most likely the Posta Medicea, built at the behest of Grand Duke Ferdinando I de' Medici starting in 1584 with plans by architects Bernardo Buontalenti and Simone Genga. With its loggias, salons, and numerous rooms, the building was called 'Osteria Grossa' for centuries and was used as an inn and horse-changing station. Over the years, many notable travelers stayed there, Charles Dickens included. However, Amoretti had a very different opinion from Dickens: during his personal Grand Tour of France and Italy in the mid-1840s, the English writer called the Osteria Grossa 'a ghostly, goblin inn … full of such rambling corridors, and gaunt rooms, that all the murdering and phantom tales that ever were written, might have originated in that one house'.[73]

Amoretti did not have the opportunity to taste any dishes at the Osteria Grossa because he dined at the town's Capuchin monastery, where he was able to enjoy 'a good dinner, and among other things an excellent tench from Lake Bolsena'. Before dinner, November darkness having already descended, moving only with the help of a local guide and the moonlight, Amoretti had climbed to the abandoned fortress looming over the town to view a rocky mass of 'dense lava crystalized into blades' he had read about in a travel book. A taste for rustic comfort and an interest in rural landscapes thus went hand in hand with the naturalist curiosity that would lead him a few years later to be a counselor to the Mines for Napoleon Bonaparte and Eugène de Beauharnais.[74]

Amoretti's experience in central Italy did not differ greatly in its elements from today's rural tourism: good quality local food and drink; contact with crops, small rural towns, and nature; and typical accommodations, whether seedy inns or sixteenth-century country mansions.

Only three years earlier, Amoretti had found himself in roughly similar circumstances in the Alps of northern Piedmont in late July and early August 1798, visiting the gold mines of Prince Borromeo. The Borromeo family had held rights to the mines in the Anzasca Valley since the second half of the fifteenth century—granted by the Sforza family then dukes of Milan—and generally contracted out their management. Amoretti had arranged with Marquis Cusani, Prince Borromeo's son-in-law, to gain access to the mines and give advice along with other experts. The company included four other men in addition to Cusani and Amoretti. Of note among these were the architect Giuseppe Zanoia, a professor at the Brera Academy of Fine Arts and designer of many buildings in Milan, and the businessman Giovanni Battista Simonetta, who traded in various products between Switzerland and northern Italy. At the end of the journey, they also planned to reach the slopes of Monte Rosa at the end of the Anzasca Valley and climb up to see the glacier.[75]

The company entered the Ossola Valley—from which a number of side valleys branched off, including the Anzasca Valley with the Borromeo mines—and reached Lake Margozzo on foot while their luggage was carried by boat up a section of the canal that flowed out of the lake. At Margozzo, the travelers found horses for themselves and a mule for their luggage, while a man took charge of transporting wine and poultry. The group carried 'a good provision of salted meats, live chickens, chocolates, coffee, liqueur, and wines'. They were also accompanied by a servant 'fit to be a cook', and Amoretti had his own personal servant (and diviner, suggesting that the ramparts of science were not entirely impermeable), a certain Vincenzo Anfossi. The group passed among farmhouses and through villages, stopping to eat and interacting with locals to set up their scientific equipment. For example, in the village of Vogogna, Amoretti asked a blacksmith to repair his geologist's hammer. Although the valley floor featured a number of vineyards, meadows, and arbors, the setting was generally quite wild, with 'precipitous peaks' looming. However, the trip through the Anzasca Valley to Monte Rosa was cut short on August 1 because of 'threatened aggression' probably due to the political instability of the period.[76]

So far we have analyzed food and lodging facilities in three very different types of environments: Scandinavian forest, central Italian countryside, and

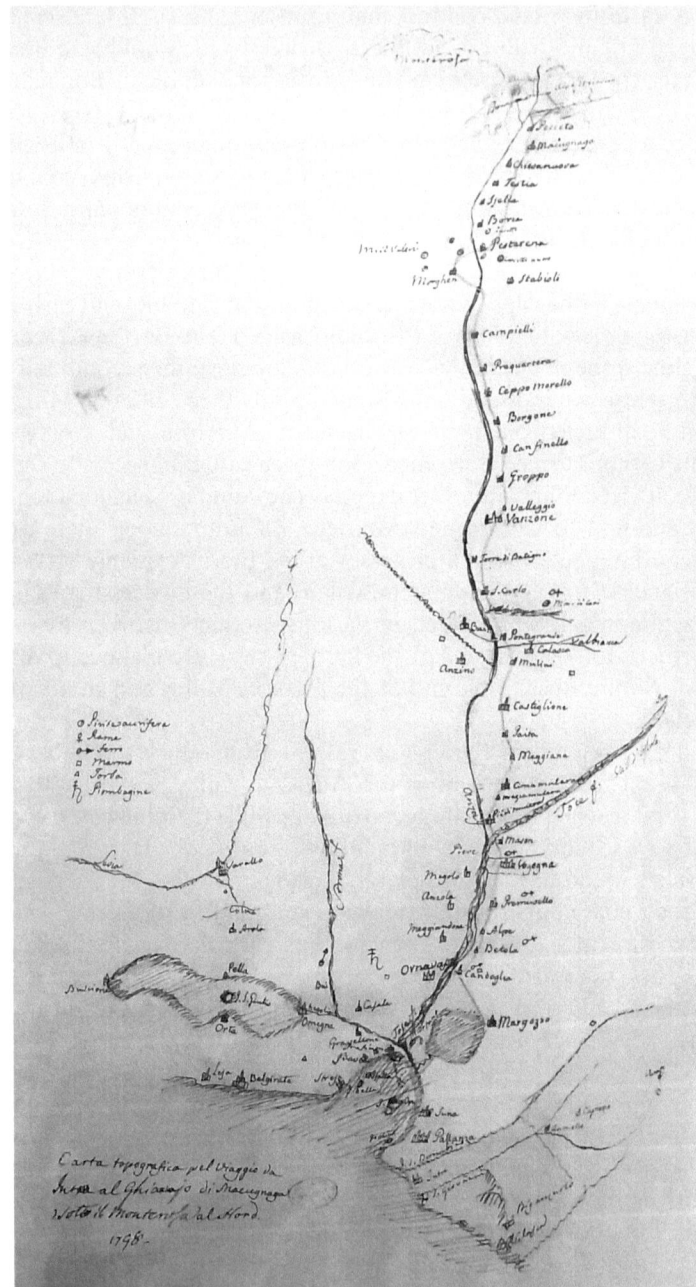

Figure 3.3 Map of the Ossola Valley and the Anzasca Valley with mines of gold pyrite (*pirite aurifera*), copper (*rame*), iron (*ferro*), and lead (*piombagine*), marble quarries (*marmo*), and peat bogs (*torba*). Istituto Lombardo – Accademia di Scienze e Lettere (Milan, Italy), *Archivio manoscritti*, 18, notebook IV, p. 300.

Alpine valley. A fourth naturalist offers information on an entirely different context: we return to Giambattista Brocchi and his journey into the Egyptian hinterland. When he set out on the Nile in late 1822 on the feluccas we mentioned in the previous section, Brocchi's company carried provisions for five months. There was hardtack, rice, barley, salt, oil, and vinegar for the company generally, but each traveler could bring his own food. Brocchi therefore armed himself with 'a box full of delicacies': almonds, pine nuts, raisins, dried figs, hazelnuts, sugar, and chocolate.[77]

In any case, the company along the way had a chance to pass through the lushly cultivated lands of the Nile Valley and certainly did not lack opportunities to buy fresh food at the various villages. Some of Brocchi's descriptions at least give insight into the potential of the land around them. For example, at the end of January 1823, they landed at a village to make an excursion on shore and then take to the river again in Abutig. Although the detour was made on donkeys without tack for a ride that was 'doubtlessly most uncomfortable', they saw thick fields of young wheat and barley, broad beans and rapeseeds in bloom. There were meadows of clover and fenugreek on which numerous cows, sheep, goats, donkeys, and dromedaries grazed. On other occasions, the company also stocked up on fish from fishermen. At an earlier stop, in Assiut, Brocchi had met a prominent Italian physician, Paolo Anino, who was under contract to produce date brandy and Greek civet, so we can imagine that he offered some to Brocchi in a gesture of hospitality.[78]

Even after they entered the desert, the travelers could find new sources of food. For example, Brocchi described the foods of some of the nomadic shepherd groups with whom he came in contact, such as certain wheat flatbreads cooked in ashes (they supplied themselves with wheat either in the Nile villages or at El Qoseir) and the milk from their flocks. Once they arrived in El Qoseir, guests of the governor, Brocchi was able to ascertain that the Red Sea in that area gave good fish, while he did not get the chance to eat any of the tasty shellfish.[79] Regarding food, Brocchi had no particular problems along the Nile or even in the drier areas. Moreover, it is understood how the mining expedition commissioned by the *Wali* was important, and thus, the supplies were carefully planned.

As for the comfort of the overnight sleeping arrangements, we have already described in part Brocchi's discomfort in the felucca, in a cramped space, poorly sheltered and subject to temperature swings. However, we also noted that the experience allowed him to enjoy nature on the Nile more. When Brocchi made his reconnaissance of the lead and galena deposits in the valleys near El Qoseir in February, the caravan he was part of was very large: ten people including himself made up the *Wali*'s group from Cairo; they were joined by five miners, the sheikh in charge of the district where the deposits to be visited were located, six of his soldiers and 'various people in tow'; there were fifty dromedaries for moving and transporting them, plus a donkey to enable Brocchi to do reconnaissance alone. In the arid, uninhabited valleys that the company visited, the only lodging was, of course, tents, which were pitched at each stage.[80]

Let us now advance sixty years to the next example. *Recollections of a Happy Life* offers interesting information about food and lodging during Marianne North's travels, just as it has regarding her perception of nature and means of transportation. For example, when she was in South Africa, she had the opportunity to stop at several facilities run by settlers. One of these was a large farm called Groote Post (written Groote Port by North), not far from Malmesbury, 'a most comfortable old place' run with an iron fist by Miss Duckitt (spelled Duckett by North), 'a regular Queen Bess or Boadicea for ruling men'. The woman was in fact filling in for her brother who was temporarily in England. Through acquaintances, she learned that North was visiting Cape Town and invited her to stay with her for several days in September 1882. To give an idea of the business Duckitt had to run, North recorded, 'Every morning she gave out over 100 rations of bread, meat, spirits, etc. Every morning a sheep was killed, and every week a bullock'. In addition, North learned that when the woman discovered that an outbreak of smallpox had occurred at a mission station three miles away some time earlier, she quarantined the farm and took it upon herself to vaccinate all the workers, with the help of a niece, preventing disaster.[81]

The Groote Post setting is particularly interesting for two reasons. First, it still exists under different ownership and boasts internationally recognized excellence in dairy farming and fine wine production. It has a restaurant, and it is possible to take tours of the large property that is home to a wide variety of flora and fauna. It is therefore a setting where nature tourism and rural tourism are well combined. The second interesting aspect is that Miss Duckett was Hildagonda Duckitt (1840–1905), an important food writer who achieved fame a few years after North's visit and whose recipes inspire today's restaurant at Groote Post. In addition to experimenting at the stove with English and Boer recipes and taking an active part in running the family business, Duckitt was also a party-thrower, and Groote Post teemed with guests, dances, concerts, and the like. An atmosphere that North got to appreciate firsthand.[82]

Needless to say, the numerous servants at the Duckitts were native Africans, of whom North noted—using the terminology of the time—that there were representatives of many ethnic groups, for example, the local Khoi, some people native to Namaqualand and others more generically from East Africa. There were also some Malaysians. However, Duckitt also had European helpers, for example, an elderly Irishman who managed fifty ostriches on the property; their feathers were sold for '£8 for one good bird, and under £2 apiece for the others all round'. Groote Post also produced a good quantity of candles (made from the fat of rams), raisins, wine, grape-jam, and butter. In the case of butter, since there was no market in which to sell it regularly, Duckitt salted and stored it for the times when there was a shortage in Cape Town, at which point she removed the salt and proceeded with the sales.[83]

When North arrived at Groote Post a week early, she found that the rooms on the second floor were still occupied by eight Boer ladies. They were quite willing to make room for North, even letting her have the best room, and soon the ladies were bonding with each other, even taking picnics and out-ings in the surrounding area. At other times, North devoted herself to her paintings of animals and plants. The rooms were spacious and pleasant, well kept, and decorated with fresh flowers. As for food, everything eaten there was produced on the farm, including a very strong wine that reminded North of Madeira. In addition to the livestock, there was also an extensive fruit and vegetable garden, a couple of date palms, pomegranates, and loquats from Japan. Among the foods enjoyed there were rice dishes, kabobs, cakes, and apple pies. There was also decent coffee, butter, dried fruits, and grape jam. North found it 'most delicious'.[84]

The next leg of the South African trip was completely different. North stayed in Stellenbosch, 'a fine old Dutch town in the midst of gardens hedged with pink and white roses, beautifully tidy, and surrounded by finely formed mountains'. The town was very peaceful, in a setting where man-made nature had as its backdrop the vast open spaces of South Africa: in addition to roads edged everywhere with roses, the surrounding countryside was dotted with wild pears, apples, and umbrella pines, while 'cypress-trees gave it quite an Italian look'. North stayed in a boarding-house kept by an old Boer couple, in which a 'beautifully clean' room was reserved for her.[85]

At first, the tranquility of Stellenbosch, the cleanliness of the lodging, and the pleasantness of the landscape made North think of extending her stay to a week, but she soon had to reconsider. Everywhere in Stellenbosch the air was filled with the psalms sung in the many schools in the surrounding area, part of a network run by the powerful local Reformed Church: 'it was most doleful', North lamented. Moreover, the residents—some of whom were familiar with her paintings—were rather intrusive in their friendliness by the painter's standards. 'I breathed more easily when I escaped from that Calvin-istic settlement'. Thus ended Marianne North's brief experience in Stellenbosch.[86]

On October 10, she arrived at the station in Tulbagh, another town immersed in the South African wilderness (the inhabitants were all Boers, except for the English district doctor). North described the mountains that bordered the wide valley as 'very grand, with perpendicular precipices and the boldest outlines', while the countryside was cultivated in a manner similar to what she had already seen in Stellenbosch. She paid 6 shillings a day for room and board (including wine) in the house where she stayed; the environment was quiet and pleasant, ideal for painting, and North stayed fifteen days. She could easily move around the surrounding countryside by train or cart, thus looking for the plants she wished to paint.[87]

We have therefore analyzed three very different lodging experiences that North had in September–October 1882, although these experiences were all in the same area and chronologically close to each other. In the case of Groote Post, this was by all rights a rural tourism experience, in which we recognize patterns very similar to those that the farm still offers its visitors today.[88] But we must keep in mind that North was in South Africa to work on her scientific paintings. In her stays in Stellenbosch and Tulbagh, on the other hand, we have seen two instances of urban accommodation—one negative, one positive—that North referenced in her writings.

Equally interesting are North's accounts of the presence of different cultures: English, her own, and Boer culture. At times, they were well amalgamated; at other times, there was a certain amount of conflict, perhaps exacerbated by the Boer victory in the First Boer War against the British fought from late 1880 to early 1881. Most of the places visited by North, although of 'Dutch' origin as she often emphasized, were part of Cape Colony, which had been British since the early nineteenth century. In contrast, the First Boer War had concerned relations between Cape Colony and the neighboring independent Transvaal Republic (Boer), thus a framework that at first glance might seem different. However, we may discern a veiled bias on North's part against the Boers and their descendants rooted in the troubled history of South Africa in the nineteenth century. Moreover, a background theme in North's accounts is the presence of indigenous peoples, and Malaysians, in a subservient status to the 'whites'.[89]

We close this Section 3.3 with a comparison of two very different case studies, despite the fact that both feature two North Italian Catholic priests as protagonists. In the first case, food and lodging for the travelers were not provided by Westerners. This case regards the Catholic priest Giovanni Battista Balangero, who was in Sri Lanka from 1874 to 1885 as a missionary after a few years spent in Australia.[90]

A keen student of both nature and local society, Balangero was particularly interested in Buddhism, its similarities and differences with Christianity, and its implications on the customs of Sinhalese society. In his accounts and interpretations in the book that chronicled his thirteen-year mission between the Pacific and Indian Oceans, he also described the influence of Buddhism on practical aspects of daily life, such as supporting travelers in rural Sri Lanka. In this regard he wrote:

Traveling in the villages of Ceylon, one often encounters, along the roads, shelter houses and canopies for travelers, where those who are tired can rest in the shade and coolness; and then also from time to time one sees stone jars containing water to drink, which people take care to keep full at all times, so that thirsty travelers can drink on their way.[91]

It was thus a real support network for the traveler, simple and certainly not aimed at making money but nevertheless very effective in its management by ordinary people who applied Buddhist moral precepts to the practical aspects of life. Buddhism is still the most practiced religion in Sri Lanka today, and this aspect is an intrinsic part of today's traveler experience in that nation. Balangero was in Sri Lanka with the serious task of Catholic missionary work, certainly not to enjoy vacation time in a setting away from home. So his study of the traveler support network and his descriptions of Buddhist temples, ceremonies, and festivals, as well as their sacred connection to natural elements—such as rivers—were part of a cultural and anthropological study that was a prerequisite to his employment.[92] However, it is interesting to note that similar elements are now considered as pathways of cultural tourism and religious tourism in Sri Lanka and regarded as such by both websites and scholarly literature.[93]

As a foil to Balangero's experience in Ceylon, we shall reference the accounts of travels by Italian priest Luigi Marchelli in northern and central Italy in the 1860s and 1870s. Marchelli left a great deal of information in his notebooks about eating habits during his tourist travels, during which he visited both rural locations and urban parks and, more generally, admired the fusion of the natural and the anthropic, as we saw in Chapter 2.[94]

What Marchelli found in Italian hotels was a highly varied gastronomic offering. For instance, in September 1865, he ate at the Albergo del Commercio in Genoa.[95] On the 4th, he wrote: 'we returned [he was traveling with his father and brother] to the hotel and had lunch in the adjoining trattoria. We were not dissatisfied with the food served to us or with the soup, we found everything good and well made'. On the 5th: 'good soup and two dishes among which was a large and very fine fish'. On the 7th: 'bread, soup, beef, schnitzel and roast pigeon with salad'. We also have materials from an 1867 trip of his. A receipt dated September 19 from the Albergo d'Italia located in Varallo, a town set in the woods of the Alpine valley Valsesia, reports soup, boiled meat, chops, trout, bread, white wine. On August 25, 1868, in Bergamo he ate with his father and brother 'a soup of pasta, stewed veal with potatoes, and fruits, with meal wine'. On August 26, they consumed 'rice and cabbage soup, beef with potatoes and stewed veal', fruits, cheese, and wine. In September 1876, in Parma he ate soup, meat with mushrooms and tomatoes, gorgonzola cheese, and wine.

In general, Marchelli's travels show a good taste for local and genuine foods, often accompanied by the positive feelings experienced in observing the environments visited: not only interesting buildings but also rural landscapes and urban parks.

3.4 Local People

We have discussed how important it was for foreign travelers to have contact with locals and benefit from their knowledge of the area. We have seen evidence of this when travelers such as Giuseppe Acerbi or Mary Kingsley found themselves aboard boats suited to navigating particular waterways. Marianne North availed herself of local contacts to move between islands in the Seychelles and when she lodged in various rural locations in South Africa. Giambattista Brocchi and Anna Maria Walker needed local support when they traversed the Egyptian desert and the Ceylonese wilderness, respectively, over rough terrain, as did Carlo Amoretti as he explored Alpine valleys or the relatively quiet countryside of central Italy. Mary Wollstonecraft Shelley, when visiting the Alps and the glaciers in Savoy in the summer of 1816 (Chapter 1), relied on locals who knew how to negotiate these zones in safety, noting that 'the men serve for guides in the summer which is lucrative', thus suggesting that it was seasonal work for them.[96]

It was not just a matter of having the right means of transportation available, perfected by long use on specific waterways or terrain. It was also important to find those who knew how to handle it and who knew the quickest and safest route to take. The same can be said for those who offered accommodation both as a business and in the spirit of hospitality.

This last section of Chapter 3 serves to analyze the role of local people in nature travel in the eighteenth and nineteenth centuries. While they are mentioned in the past sections in relation to the various topics analyzed—means of transportation, road network, hospitality—in the following pages we will devote our attention exclusively to these people, focusing on their profiles and the dynamics of their interaction with foreign travelers. We will thus try to give greater depth to figures who always appear in the background of travel accounts but who played an indispensable role with their in-depth knowledge of local nature.

Our first accounts in this section are from Richard Pococke (1704–1765), an Anglican bishop of the Church of Ireland who visited parts of Europe, Egypt, the Levant, Asia Minor, and Greece on several trips during the 1730s and early 1740s. These travel experiences were not motivated by pastoral care or missionary goals, as has been the case for other clergymen analyzed in this volume. Pococke's travels fell into the category of the Grand Tour—he came from a wealthy family—and do not fit our definition of 'incidental tourism', that is, travels undertaken for other purposes such as scientific research or work that led to contact with nature. However, in a series of tours and pastoral visitations that took place between the late 1740s and the 1760s, Pococke also explored Ireland and Scotland.[97]

In his travel accounts—some published while he was still alive, others posthumously—there are accounts of interaction with locals in the rural context that provide an excellent introduction to the case studies in this Section 3.4. For example, from June to October 1752, Pococke made a tour along the coast of Ireland, taking an interest in the natural aspects, history, culture, and society. The starting and finishing point was Dublin: Pococke's route went counterclockwise and touched almost all the counties that bordered the sea, cutting inland from Limerick to Cork. He traveled on horseback and through areas of Ireland that were far from safe and comfortable to cross, against a backdrop of enduring poverty.[98]

However, in the towns and villages he visited, he was hosted both by the local gentry and by more modest farmers, who provided him not only with rooms in their dwellings in which he could sleep but also with company, food, and wine. Typically, this was simple hospitality, in which the food consisted of the products of local agriculture, livestock farming, or possibly fishing, but Pococke was never disappointed. On other occasions, however, the use of guides and the help of local individuals served to move more safely and comfortably through the Irish countryside.[99]

The travel notes taken by Pococke in 1752 thus give interesting insights not only into the role of locals in facilitating transit through the Irish countryside but also in making the travel experience more immersive. In this respect, Giovanni Battista Brocchi's *Giornale delle osservazioni* also provides an abundance of information. Indeed, the Italian geologist was generous in crediting local guides and hosts with supporting the success of his travels between 1822 and 1826.

For example, we mentioned the composition of Brocchi's expedition to study the deposits near El Qoseir. Let us now dwell on the criteria used by the Egyptian government to choose his supporting figures. Brocchi, the chemist Giuseppe Forni, and their collaborators were paired with an interpreter who spoke Italian, Arabic, and Turkish and who traveled with them on the main felucca so as to translate requests and responses as needed in the villages along the Nile and then in the inland crossing from Qena to El Qoseir. Twelve miners traveled on the second felucca, with the purpose of intervening once they reached the deposits at the end of the voyage and drilling cores as Brocchi requested. Interestingly, the government chose Greek and Armenian miners because they were Christian and not Muslim, believing that this would result in their greater obedience to the group of surveyors led by Brocchi. In addition to Forni, in fact, Italian and Swiss experts were part of the expedition, and we have reason to believe they were all Christians, at least nominally.[100]

Experts, miners, and the interpreter were accompanied by a third felucca with five soldiers and three officers. The stated role of the soldiers was to escort the expedition, give the necessary orders to Qena to set up the caravan that would cross the desert to El Qoseir, and in general 'so that throughout the journey we would be given what we needed'. Although Brocchi did not mention it, we can conclude that the soldiers were also meant to control the movements and actions of foreign surveyors on Egyptian soil, as well as to represent the Cairo government once work began on the lead and galena deposits in the El Qoseir valleys. Likewise, at that very last stage of the expedition, the sheikh of the district where the deposits were located was present. We thus realize the importance of Brocchi's mission within the Egyptian economic empowerment policies implemented by *Wali* Mehmet Ali.[101]

A different situation occurred in early September 1823, when Brocchi was in Jaffa, on the coast of present-day Israel. As we considered in Chapter 1 for this specific case, but as we have also seen for travelers in other areas of the globe, the context was the much less difficult one of nature domesticated by humans, that is, crops in anthropized areas or natural scenery accurately reproduced in pleasure gardens or botanical gardens. However, even in Jaffa, interaction with locals proved essential for full enjoyment of the land.

Passing through lush fields, orchards and vineyards, beautiful gardens and tree-lined avenues, Brocchi arrived in the city and fortuitously met the Austrian consular agent in Jaffa, Francesco Damiani, a diplomat born in Jerusalem to a French family. Brocchi had never met him before and was unaware of his identity, initially avoiding him due to his 'grotesque' and apparently eccentric attire. Having resolved the misunderstanding, Brocchi found Damiani to be an excellent host. He invited Brocchi into his home, offered him coffee and brandy, guided him through the streets and parks of Jaffa, kept him company in the evenings, and even hosted him a few months later when, in April 1824, Brocchi was on his way back to Egypt. In Damiani's case, the guidance of a local personality facilitated Brocchi's visit in a context where the flora was artificial, however much relief it provided after the aridity of the desert. Today we would say that the dividing line between urban tourism and rural tourism was very blurred during Brocchi's stay in Jaffa.[102]

A few weeks later, toward the end of September, Brocchi was instead on Mount Lebanon, and in this case, the equally fortuitous presence of another diplomat gives us evidence of rural tourism intertwined with a more authentic nature tourism. Indeed, after inspecting a natural oil shale deposit, Brocchi and his local guides and escort soldiers crossed paths with the French consul in Tripoli, who was riding free and alone through the countryside. He explained to Brocchi that he had come to spend a few weeks in a nearby village. Indeed, the area lent itself to a full immersion in nature as a break from consulate duties, including shady valleys, forests, pleasant streams, and the nearby Maronite monastery of Qozhaya, rich and offering excellent hospitality (we do not know if the French consul passed through there during his vacation, Brocchi certainly visited). In the area there was also a hermitage dedicated to St. Sergius where only an Italian Carmelite lived: there was a beautiful view of the area from the hermitage, so much so that a previous French consul in Tripoli used to spend his summers nearby.[103]

Interestingly, those valleys were an area traditionally used for vacationing by French diplomats from Tripoli, a seven- to eight-hour walk away. As for attention to visitors, the monks of Qozhaya were accustomed to hospitality due to the presence of many pilgrims to the attached shrine. In addition, there was also a convent run by nuns offering hospitality exclusively to women. As for Brocchi's armed escort and the Turkish style of the clothes worn by the travelers, the consul admitted that at first glance they had alarmed him, believing that the company had no good intentions toward him despite the fact that they were in the midst of the Ottoman eyalet of Tripoli.[104]

Changing latitude and climate, U.S. explorer Charles Francis Hall experienced the nature and culture of the Inuit in the Arctic Circle, in a very different context from Brocchi's experience in Lebanon. Inuit guides played an important role in Hall's expeditions. For example, for the 1860–1862 expedition, during which he explored Baffin Island, the assistance of a pair of Inuit guides and interpreters, Ipirvik and Taqulittuq, was indispensable. The main objective of the expedition was to find evidence in Inuit oral tradition of the explorations of English navigator Martin Frobisher in the 1570s and John Franklin's British expedition, which set out in 1845 and completely vanished.[105]

The comprehensive knowledge that Ipirvik, Taqulittuq, and their fellow Inuit had of the complex nature of the Baffin Islands was crucial to Hall's expedition. This was clear in January 1861, when Hall set out from the bay where the whaler *George Henry* was moored, after giving him passage from the United States, to reach Cornelius Grinnell Bay. Hall was accompanied by Ipirvik, Taqulittuq, an Inuit named Koodloo, and a varying number of other companions. The means of transportation were dog sleds. In addition to sleeping bags, blankets, and heavy clothing, the supplies the group brought with them consisted of 1.5 pounds of preserved boiled mutton in cans, 3 pounds of raw salt pork, 4 pounds of sea-bread, a quarter pound of pepper, 2 pounds of ground burnt coffee, a quarter of molasses, a quarter of corn-meal, 3 pounds of Cincinnati cracklings for soup. It was immediately clear to Hall the skill with which Ipirvik drove the sled and steered the dogs on a complicated route with many obstacles. Not surprisingly, the trip was supposed to take one day, but more time was needed because of bad weather.[106]

The importance of Inuit guides was fully manifested in Hall's eyes in the construction of igloos in which to sleep each night. Ipirvik and Koodloo knew where more ice blocks could be obtained and knew how to saw them quickly and how to arrange them. Hall's skills in this regard were obviously limited, so he helped mainly by transporting the blocks from the point of extraction to the point of igloo construction and then served as a laborer under the instructions of the two experienced Inuit. Taqulittuq, on her part, knew how to set up the interior of the igloo quickly, supervising the preparation of dinner and managing the sleeping arrangement. In addition, the group was forced into the igloo for many, many hours because of a terrible gale, which stirred the sea and caused large cracks to open in the ice floe not far from the shelter. Here, too, the knowledge the Inuit had of the local terrain and climate was essential in the step-by-step planning of the trip to their destination. As Ipirvik, Taqulittuq, and Koodloo planned the best strategies, Hall noted fearfully that 'the heavy sea kept the ice creaking, screaming, and thundering, as it actually danced to and fro! It was to me a new but fearful sight'.[107]

Figure 3.4 The igloo in the storm, January 1861. Charles Francis Hall, *Arctic Researches and Life among the Esquimaux: Being the Narrative of an Expedition in Search of Sir John Franklin, in the Years 1860, 1861, and 1862* (New York, 1865), p. 198.

Ipirvik and Taqulittuq were not without a rich background of contacts with the Anglo-Saxon world. Some of Taqulittuq's relatives were travelers and at least one of them had visited England in 1839. In 1853, Ipirvik and Taqulittuq had been received by Queen Victoria after some whalers had taken them to Britain, and they had lived for nearly two years there in good conditions, in this way deepening their knowledge of British society. Ipirvik had excellent knowledge and experience of sailing along the coast of Baffin Island and its environs and had already served as a guide in that sea to the whaling captain John Parker and his ship *Truelove*. Parker himself was very interested in Inuit society and culture and had repeatedly urged British authorities, missionaries, and public opinion to improve the living conditions of the Baffin Island Inuit. This was the setting in which Ipirvik and Taqulittuq had achieved fame among explorers and navigators in the Canadian Arctic, so much so that Hall himself had been very eager to meet them before making their acquaintance in November 1860 on the *George Henry*.[108]

Hall seemed to have a soft spot for Taqulittuq, for in more than one passage he described her with particular admiration. We quote Hall's description of their first meeting:

November 2, 1860. While intently occupied in my cabin, writing, I heard a soft, sweet voice say 'Good morning, sir.' The tone in which it was spoken—musical, lively, and varied—instantly told me that a lady of refinement was there, greeting me. I was astonished. Could I be dreaming? Was it a mistake? No! … I raised my head: a lady was indeed before me, and extending an ungloved hand.[109]

Already from this excerpt we can guess Hall's admiration. The particular aspect is that this was strongly dictated by the fact that Taqulittuq was strongly Westernized, and this aspect probably made her more comprehensible to Hall. In fact, the woman stood against the light on the threshold to his cabin and Hall could at first notice only the attire in which he was wont to see a woman: 'crinolines, heavy flounces, an attenuated toga, and an immensely expanded "kiss-me-quick" bonnet'. When he saw Taqulittuq's face, Hall was extremely surprised: 'Whence, thought I, came this civilization refinement?'[110] Hall's comment is part of the complicated contacts between Western culture and Inuit culture, in an ongoing contrast between understanding and prejudice on the part of the former toward the latter. In 1884, a few decades after Hall's experiences, German-born anthropologist Franz Boas (1858–1942) traveled to Baffin Island to live with the Inuit and study them. He was fascinated by that culture but could not fully grasp it. For example, he did not understand the smells and the sounds of their tradition, the conception of time and rhythms, and the interaction with the Arctic environment. However, his experience enabled him to orient himself toward cultural relativism, and he became convinced that Westerners and especially those who had received a higher education—a group he felt he belonged to—had no right to look down on and regard cultures other than their own as ridiculous.[111]

To return to Hall, although his surprise only increased his liking for Taqulittuq and although the man had nothing but admiration for her, her husband, and the other Inuit guides, the sense of superiority felt by the 'civilized man' seems obvious to us.

North's accounts also provide interesting examples of local figures she relied on in her many travels around the world, and many of these were important for her contact with local nature and culture. Indeed, interaction with both native individuals from the areas she visited and with immigrants and diplomats enabled her to experience full immersion in the contexts she visited. We have already had occasion to point this out in several places in the book. We add here two examples relating to the 1870s and 1880s.

When North visited parts of North America in 1871, she wanted to see the Montmorency Falls near Quebec City. She stayed in the cottage of a couple of friends and then in an inn; she hired a local boy who carried her tools and knew the area, guiding her to particularly interesting spots around the Falls. Later, she was also able to visit a mixed village of Frenchmen and Hurons. Here she purchased handicrafts in moose hair. She was guided by a young man 'with long lank hair and high cheek-bones' who gave her a tour of the school, where the Huron children sang her some pleasant traditional hymns 'with soft-sounding words' (North did not understand the language) that the teacher told her had never been put in writing.[112]

Years later, in August 1884, North set out on her last major voyage, which would last until early 1885: the destination this time was Latin America with the goal of painting some particularly rare plants. The voyage from England stopped in Bordeaux, Rio de Janeiro, and Montevideo and then sailed south to the Strait of Magellan. It then continued up the Pacific coast of Latin America to Lota, where in the few hours ashore North visited the exotic park commissioned by the rich and powerful Luis and Isidora Cousiño, of the coal-mining magnate family. Then the ship continued on to Concepción and Valparaiso. At that point, North proceeded inland in Chile to Santiago, taking advantage of the railroad line that had been opened in 1863 after more than a decade of state-funded work slowed by the difficulties that the mountainous terrain posed to the undertaking.[113]

Santiago would serve North as a base for several trips into the wild (but also some social life) until the new year. At that point she set sail again for Britain, quite fatigued, this time via Lima and Panama, where she saw the first attempts to build the canal. She made a final stop in Jamaica before returning to Europe.[114]

Beyond the circle of British diplomats and local notables she frequented in the Chilean capital, she got to enjoy rustic hospitality when she resided for more than two weeks in the town of Apoquindo, located in the countryside near Santiago. North went there in search of some special flowers to paint. Apoquindo was already a renowned hot springs resort frequented by Santiaguinos, but she was there in the off-season and the spa was almost empty. She thus had her pick of rooms and received 'every sort of kind care from the people there', although she took no interest in spa tourism and devoted herself to exploring the flora in the surrounding area. Wanting then to see some beautiful blue puyas that were at a higher altitude, she hired a guide and procured horses with which to ascend into the mountains. The two left the mounts when the road became too steep and proceeded on foot in the clouds. When these thinned out at a higher altitude, North could enjoy the beauty of the puyas, 'noble flowers', against the backdrop of a snowy peak and a clear sky.[115]

North's circle of acquaintances in Santiago included Benjamín Vicuña Mackenna, Chilean writer, senator, and president of the Partido Liberal Democrático. He greatly admired North's paintings, so much so that he invited her and an English friend to spend a few days at his country house. The hacienda consisted of a single-story wooden main building surrounded by a beautiful veranda and garden; there were also stables and other facilities for rural work. There were cattle on the hacienda owned by two young and wealthy partners of Vicuña Mackenna, to whom he provided 'land and food' and a share of profits. The place was also used at certain times of the year by Vicuña Mackenna's wife and other relatives for vacation.[116]

As Vicuña Mackenna was a prominent and busy political figure, the hacienda was under the direction of a manager, who lived there with his family. Vicuña Mackenna and North's English friend returned to Santiago after two days, while she stayed for more than a week. The manager's wife and mother-in-law acted as guides to the painter in the surrounding countryside to find the flowers and plants she wished to depict, but only a few interesting specimens were found. North was much more impressed by the rodeo they took her to see in a sort of amphitheater in the hills where 300 horsemen 'in flaming ponchos and hats as big as targets' were gathered. She described the energy of the crowd and the wild action of the rodeo, commenting on the fusion of local rural folklore and the natural background: 'The dust and the sunshine and the general excitement were marvellous and most picturesque, surrounded as it all was by green trees and mountain-tops'. There was also music, dancing, food, and 'a huge jag of beer was handed round, and every one had to take a sip from it', but North and the other women were escorted back to the hacienda by the manager: 'He said the manners of the company would not improve toward the end of the day'.[117]

With Marianne North's testimony, we close this chapter about the importance of local culture in forms of tourism centered on unspoiled or domesticated nature. Knowledge of the environment and its dangers and beauties, the suitability or unsuitability of various means of transportation, and hospitality as an experience of local culture and tradition are all elements of the human–environment interaction we examined earlier, a key feature of the types of tourism at the center of our book.

Notes

1　Martin Mowforth and Ian Munt, *Tourism and Sustainability: New Tourism in the Third World* (London, 1998); Joe Peters, 'Sharing National Park Entrance Fees: Forging New Partnerships in Madagascar', *Society and Natural Resources*, Volume 11, No. 5 (1998), pp. 517–530; Duncan Tyler and J. Mark Dangerfield, 'Ecosystem Tourism: A Resource-Based Philosophy for Ecotourism', *Journal of Sustainable Tourism*, Volume 7, No. 2 (1999), pp. 146–158.

2　Alejandra Echeverri, Jeffrey R. Smith, Dylan MacArthur-Waltz, Gretchen C. Daily, 'Biodiversity and Infrastructure Interact to Drive Tourism to and Within Costa Rica', *PNAS*, Volume 119, No. 11 (March 2022), https://doi.org/10.1073/pnas.2107662119. For insights into these issues, see also Amanda L. Stronza, Carter A. Hunt, Lee A. Fitzgerald, 'Ecotourism for Conservation?', *Annual Review of Environment and Resources*, Volume 44 (2019), pp. 229–253; Carter A. Hunt, William H. Durham, Laura Driscoll, Martha Honey, 'Can Ecotourism Deliver Real Economic, Social, and Environmental Benefits? A Study of the Osa Peninsula, Costa Rica', *Journal of Sustainable Tourism*, Volume 23, No. 3 (2015), pp. 339–357; Robin Naidoo and Wiktor L. Adamowicz, 'Biodiversity and Nature-Based Tourism at Forest Reserves in Uganda', *Environment and Development Economics*, Volume 10, No. 2 (May 2005), pp. 159–178; David S. Wilkie and Julia Carpenter, 'Can Nature Tourism Help Finance Protected Areas in the Congo Basin?', *Oryx*, Volume 33, No. 4 (October 1999), pp. 22–31.

3　Harold J. Cook, *Risking private ventures: the instructive failure of a well-traveled artist, Cornelis de Bruyn*, in Fokko Jan Dijksterhuis (ed), *Regulating Knowledge in an Entangled World* (Abingdon and New York, 2022), pp. 169–192; Geralda Jurriaans-Helle, *Cornelis De Bruijn: voyages from Rome to Jerusalem and from Moscow to Batavia* (Amsterdam, 1998); Jan de Hond, *Cornelis de Bruijn (1652–1726/27): a Dutch painter in the East*, in Geert Jan van Gelder, Ed de Moor (eds), *Eastward Bound: Dutch Ventures and Adventures in the Middle East* (Amsterdam and Atlanta, 1994), pp. 51–80. See also Benjamin Schmidt, *Inventing Exoticism: Geography, Globalism, and Europe's Early Modern World* (Philadelphia, 2015).

4　For chronological details of the voyage, see Cornelis De Bruijn, *Travels into Muscovy, Persia, and Part of the East-Indies: Volume I* (London, 1737), pp. 72–73, 86. If De Bruijn left Moscow on April 22, the departure from the Volga port is recorded two days later, on the 247 precisely. The quotation is taken from p. 73.

5　De Bruijn, *Travels into Muscovy: Volume I*, pp. 73–86.

6　Ibid., pp. 73, 78–79, 82.

7　Ibid., p. 72. For today's example, we refer to this Volga cruise program: *Volga Dream – Russian River Cruises: Moscow to Astrakhan 14 Days - 13 Nights*, https://www.volgadream.com/cruises/moscow-to-astrakhan/, accessed on November 14, 2022. See also Alan D. Roe, *Into Russian Nature: Tourism, Environmental*

Protection, and National Parks in the Twentieth Century (Oxford and New York, 2020); Lidia Andrades and Frederic Dimanche, 'Destination competitiveness and tourism development in Russia: Issues and challenges', *Tourism Management*, Volume 62 (October 2017), pp. 360–376. For the impact of the war on international tourism in 2022, refer to the UNWTO reports published in *Impact of the Russian Offensive in Ukraine on International Tourism*, https://www.unwto.org/impact-russian-offensive-in-ukraine-on-tourism#:~:text=A%20prolonged %20conflict%20could%20translate,US%24%204.7%20billion%2C%20respectively, accessed on November 14, 2022.

8 The two quotes are taken from Giuseppe Acerbi, *Travels through Sweden, Finland, and Lapland, to the North Cape, in the Years 1798 and 1799: Volume I* (London, 1802), pp. 373, 375.

9 Ibid., pp. 373–375.

10 Ibid., p. 374. Acerbi noted that, precisely because of the difficulty of ascending the Tornio at that point, the locals tended to proceed overland, despite the difficult conditions of travel through thickets and marshes. They had carved out a path by felling trees, clearing the ground, and placing logs on the ground at the most dangerous points (p. 375).

11 Ibid., pp. 371–372.

12 *Kattilakoski Rapids on the Tornio / Torne River*, in the website of Pello, Lapland: https://travelpello.fi/en/attraction/kattilakoski-rapids-tornio-torne-river-lapland-finland/, accessed on November 14, 2022; *Torne River Rafting*, in the website of Icehotel, https://www.icehotel.com/torne-river-rafting, accessed on November 14, 2022; Ruunaa Hiking Park, in the website nationalparks.it, https://www.nationalparks.fi/ruunaa, accessed on November 14, 2022.

13 BCT, *Fondo Giuseppe Acerbi, Carte*, IX, 2, notebook 3, pp. 6–7.

14 Among the articles and web pages describing today's sailing on the Nile, we point to these examples: Annamaria Giannetto Pini, *In feluca sul Nilo: una crociera alternativa*, https://www.viaggionelmondo.net/25350-in-feluca-sul-nilo-crociera-alternativa/m accessed on November 14, 2022; Emma Thomson, *Life on the longest river: sailing 140 miles along the Nile*, August 5, 2019, https://www.nationalgeographic.co.uk/travel/2019/07/egypt-life-longest-river, accessed on November 14, 2022. See also Abdelazim M. Negm, *The Nile River* (Cham, 2017).

15 Giovanni Battista Brocchi, *Giornale delle osservazioni fatte ne' viaggi in Egitto, nella Siria e nella Nubia: Volume I* (Bassano del Grappa, 1841), pp. 233, 297, 328. Brocchi indicates as the destination of the Nile journey *Kenneh*, a river town from which he would then cross the desert to reach the mines at El Qoseir, on the Red Sea, which we mentioned previously in this book. Both because of the geographical location (five-day caravan from El Qoseir) and the assonance of the name (in ancient Latin it was *Caena*), we conclude that Kenneh was indeed Qena.

16 Brocchi, *Giornale delle osservazioni: Volume I*, pp. 234–235.

17 Ibid., pp. 264, 284.

18 Ibid., pp. 282–283 (circle), 308–309 (eclipse), 315–316 (crocodile).

19 Alison Blunt, *Travel, Gender, and Imperialism: Mary Kingsley and West Africa* (New York, 1994); Dea Birkett, *Mary Kingsley: Imperial Adventuress* (London, 1992); Valerie Grosvenor Myer, *A Victorian Lady in Africa: The Story of Mary Kingsley* (London, 1989); Katherine Frank, *A Voyager Out: The Life of Mary Kingsley* (New York, 1986).

20 For coordinates of the journey from Cape Lopez to Talagouga, see Mary Kingsley, *Travels in West Africa, Congo Français, Corisco and Cameroons* (London, 1897), pp. 105, 109–110, 123–131, 150–151. On Gabon's history, society, and culture in the nineteenth and twentieth centuries, see Florence Bernault,

Colonial Transactions: Imaginaries, Bodies, and Histories in Gabon (Durham NC, 2019); Michael Charles Reed and James Franklin Barnes (eds), *Culture, Ecology, and Politics in Gabon's Rainforest* (Lewiston, 2003); Christopher J. Gray, *Colonial Rule and Crisis in Equatorial Africa: Southern Gabon, c. 1850–1940* (Rochester NY, 2002).

21 Kingsley, *Travels in West Africa*, pp. 139–143.
22 Ibid., pp. 165–166.
23 Ibid., p. 166.
24 Ibid., pp. 168–184.
25 Ibid., p. 187. For the two descriptions above, see p. 168 and p. 171, respectively.
26 Ibid., pp. 172–174.
27 For instance, see the Facebook page *Gabon Adventure Tours*, https://www.facebook.com/GabonAdventureTours/, accessed on November 14, 2022.
28 De Bruijn, *Travels into Muscovy: Volume I*, pp. 1–5.
29 Ibid., p. 2. Richard W. Unger, 'Dutch Ship Design in the Fifteenth and Sixteenth Centuries', *Viator: Medieval and Renaissance Studies*, Volume 4 (1973), pp. 387–411, in particular pp. 404–407.
30 De Bruijn, *Travels into Muscovy: Volume I*, p. 4. For the context of the Northern Wars during the early modern period see Robert I. Frost, *The Northern Wars. War, State and Society in Northeastern Europe, 1558–1721* (Harlow, 2000).
31 De Bruijn, *Travels into Muscovy: Volume I*, p. 5. De Bruijn used the name 'New Dwinko', but from the context and description, it is possible to identify it with Novodvinskaya Fortress. See also Georgiï Lappo, *Arkhangel'sk*, in Mark Nuttall (ed), *Encyclopedia of the Arctic: Volumes 1, 2 and 3* (New York and Abingdon, 2005), pp. 149–150.
32 De Bruijn, *Travels into Muscovy: Volume I*, p. 1.
33 William Wilson (ed), *A Missionary Voyage to the Southern Pacific Ocean Performed in the Years 1796, 1797, 1798 in the Ship Duff...* (London, 1799). On the London Missionary Society see Cecil Northcott, *Glorious Company: One Hundred and Fifty Years Life and Work of the London Missionary Society 1795–1945* (London, 1945).
34 Wilson (ed) *A Missionary Voyage*, pp. 4–6, 10, 12.
35 Ibid., p. 12. See also Robert Harvey, *The War of Wars: The Great European Conflict 1793–1815* (New York, 2007); François Crouzet, *Britain Ascendant: Comparative Studies in Franco-British Economic History* (Cambridge and New York, 1990); Jeremy Black, *Natural and Necessary Enemies: Anglo-French Relations in the Eighteenth Century* (Athens GA, 1986).
36 Wilson (ed), *A Missionary Voyage*, pp. 50–51.
37 See the following texts and the bibliography indicated therein: Gudrun Andersson, Jon Stobart, *Daily Lives and Daily Routines in the Long Eighteenth Century* (New York and Abingdon, 2022); Paul A. Gilje, *To Swear Like a Sailor. Maritime Culture in America, 1750–1850* (Cambridge and New York, 2016); Isaac Land, *War, Nationalism, and the British Sailor, 1750–1850* (New York, 2009); Sam Willis, *Fighting at Sea in the Eighteenth Century: The Art of Sailing Warfare* (Woodbridge, 2008).
38 Wilson (ed), *A Missionary Voyage*, pp. 27, 91–92.
39 Charles Darwin, *Journal and Remarks, 1832–1836*, in *Narrative of the Surveying Voyages of His Majesty's Ships Adventure and Beagle, between the Years 1826 and 1836, Describing their Examination of the Southern Shores of South America and the Beagle's Circumnavigation of the Globe: Volume III* (London, 1839), p. 233.

40 For instance, see *Adventure Cruise: Cape Horn & Glaciers*, https://www.swoop-patagonia.com/adventure-cruise-cape-horn-glaciers, accessed on November 15, 2022.
41 Darwin, *Journal and Remarks*, p. 237. See also James Taylor, *The Voyage of the Beagle: Darwin's Extraordinary Adventure aboard FitzRoy's Famous Survey Ship* (London, 2008).
42 Darwin, *Journal and Remarks*, pp. 237–244. On the location between Navarino and Lennox Islands, see another important voyage report: Herbert Leland Crosthwait, 'A Journey to Lake Saint Martin, Patagonia', *The Geographical Journal*, Volume 25, No. 3 (1905), pp. 286–291, in particular p. 287.
43 Darwin, *Journal and Remarks*, p. 243.
44 For instance, see Paul Stafford, *The Best Beagle Channel Tours & Cruises from Ushuaia*, August 5, 2020: https://www.travelmag.com/articles/beagle-channel-tours/, accessed on November 15, 2022. See also *5 Nights: Wildlife, Glaciers & Cape Horn*, https://www.swoop-patagonia.com/travel/cruises/wildlife-cape-horn, accessed on November 15, 2022.
45 Marianne North, *Recollections of a Happy Life: Volume II*, Janet Catherine North as Mrs. John Addington Symonds (ed) (New York and London, 1892), p. 288. North always calls the Hoads with only the initial H, but on p. 295 she betrays herself, intentionally or not.
46 North, *Recollections of a Happy Life: Volume II*, pp. 290–292. There was also a lepers' camp established on Curieuse at the turn of the 1830s, which North visited. During her visit, however, the lepers had dwindled in number, and although disfigured by illness, they seemed to her 'contented and happy' (p. 292). About it, refer also to Deryck Scarr, *Seychelles since 1770: History of a Slave and Post-Slavery Society* (London, 2000), pp. 77–78.
47 North, *Recollections of a Happy Life: Volume II*, pp. 293–294. For a framing of society and the economy in the Seychelles under British rule and for the productive and demographic recovery in the latter nineteenth century, see the aforementioned Scarr, *Seychelles since 1770*, but also William McAteer, *Hard Times in Paradise: The History of Seychelles 1827–1919* (Mahé, 2000).
48 North, *Recollections of a Happy Life: Volume II*, p. 294.
49 Ibid., pp. 288, 295.
50 Ibid., p. 295.
51 Carl Linnaeus, *Lachesis Lapponica or a Tour in Lapland: Volume I* (London, 1811), pp. 75, 81–82.
52 Ibid., pp. 76, 80–81.
53 Ibid., p. 81, 86–87.
54 Stefano Levanti and Giovanni Liva, *Viaggio di quasi tutta l'Europa colle viste del commercio dell'istruzione e della salute. Lettere di Paolo e Giacomo Greppi al padre (1777–1781)* (Milan-Cinisello Balsamo, 2006), pp. 14–58. For the framework in which the Greppy family used to move refer to the entries *Greppi, Antonio* https://www.treccani.it/enciclopedia/antonio-greppi_%28Dizionario-Biografico%29/ and *Greppi, Paolo* https://www.treccani.it/enciclopedia/paolo-greppi_%28Dizionario-Biografico%29/ written by Elena Puccinelli for the *Dizionario Biografico degli Italiani: Volume 59* (Rome, 2002), both accessed on December 4, 2022. See also Sophus A. Reinert, *The Academy of Fisticuffs: Political Economy and Commercial Society in Enlightenment Italy* (Cambridge MA and London, 2018); Elena Riva, *Da negoziante a gentiluomo. La formazione di Paolo Greppi tra commercio, finanza e diplomazia*, in *Rapporti diplomatici e scambi*

commerciali nel Mediterraneo moderno, ed. Mirella Mafrici (Salerno and Soveria Mannelli, 2004), pp. 379–444.

55 About Amoretti's professional and scholarly profile see Lucia De Frenza, *I sonnambuli delle miniere: Amoretti, Fortis, Spallanzani e il dibattito sull'elettrometria organica e minerale in Italia (1790–1816)* (Florence, 2005). See also Franco Arato, 'Carlo Amoretti e il giornalismo scientifico nella Milano di fine Settecento', *Annali della Fondazione Luigi Einaudi*, Volume 21 (1987), pp. 175–216; Renzo De Felice, *Amoretti, Carlo*, in *Dizionario Biografico degli Italiani: Volume 3* (Rome, 1961), https://www.treccani.it/enciclopedia/carlo-amoretti_%28Dizionario-Biografico%29/, accessed on December 2022.

56 AIL, *Archivio manoscritti*, 18, notebook VI, pp. 1–259. For the ascent to the crater of Vesuvius, see pp. 107–111.

57 AIL, *Archivio manoscritti*, 18, notebook VI, pp. 26–27.

58 AIL, *Archivio manoscritti*, 18, notebook VI, pp. 35, 39, 41, 54.

59 Brocchi, *Giornale delle osservazioni: Volume I*, pp. 328–329, 367–368. See also *Giornale delle osservazioni fatte ne' viaggi in Egitto, nella Siria e nella Nubia: Volume II* (Bassano del Grappa, 1841), p. 35, note a, p. 61, note a, pp. 138–139.

60 Anna Maria Walker, 'Journal of An Ascent to the Summit of Adam's Peak, Ceylon', *Companion to the Botanical Magazine*, Volume 1 (1835), pp. 3–14, specifically p. 7.

61 Ibid., p. 8.

62 North, *Recollections of a Happy Life: Volume II*, p. 286.

63 North, *Recollections of a Happy Life: Volume II*, pp. 177–178, 228–233, 236–239. The bibliography on the European expansion in both South Africa and New Zealand and on its consequences is rich and multifaceted. In this particular case, we have based our historical, political, and sociocultural framework, respectively, on Iris Berger, *South Africa in World History* (Oxford and New York, 2009); Robert Ross, *A Concise History of South Africa* (Cambridge and New York, 2nd edition); Giselle Byrnes (ed), *The New Oxford History of New Zealand* (Oxford and New York, 2009); Tom Brooking, *The History of New Zealand* (Westport CT and London, 2004).

64 North, *Recollections of a Happy Life: Volume II*, pp. 177–178.

65 About the evolution of the railway system through the nineteenth and twentieth centuries and its importance from a global perspective, see Ralf Roth and Günter Dinhobl (eds), *Across the Borders: Financing the World's Railways in the Nineteenth and Twentieth Centuries* (Aldershot and Burlington VT, 2008).

66 For some studies in this regard, see Luis Henrique de Souza, Elisabeth Kastenholz, Maria de Lourdes de Azevedo Barbosa, 'Relevant Dimensions of Tourist Experiences in Unique, Alternative Person-to-Person Accommodation—Sharing Castles, Treehouses, Windmills, Houseboats or House-Buses', *International Journal of Hospitality & Tourism Administration*, Volume 21, No. 4 (2020), pp. 390–421; Ana Bronchado, 'Nature-Based Experiences in Tree Houses: Guests' online reviews', *Tourism Review*, Volume 74, No. 3 (2019), pp. 310–326; Philip Sloan, Willy Legrand, Sonja Kinski, 'The Restorative Power of Forests: The Tree House Hotel Phenomena in Germany', *Advances in Hospitality and Leisure*, Volume 12 (2016), pp. 181–189.

67 Linnaeus, *Lachesis Lapponica: Volume I*, pp. 77–78.

68 Matti La Mela, 'Property Rights in Conflict: Wild Berry-Picking and the Nordic Tradition of *allemansrätt*', *Scandinavian Economic History Review*, Volume 62, No. 3 (2014), pp. 266–289.

69 Linnaeus, *Lachesis Lapponica: Volume I*, pp. 78–80.

70 AIL, *Archivio manoscritti*, 18, notebook VI, pp. 25–27.

71 AIL, *Archivio manoscritti*, 18, notebook VI, p. 33. About the values and the aims of the *Borghi più belli d'Italia* association see https://borghipiubelliditalia.it/club/, accessed on November 15, 2022.

72 AIL, *Archivio manoscritti*, 18, notebook VI, p. 43.

73 Charles Dickens, *Pictures from Italy* (London, 1846), p. 160. About the history of this building refer to: Carmen Borsarelli, *La fortezza di Radicofani*, in Leonardo Rombai (ed), *I Medici e lo Stato senese: 1555–1609 storia e territorio* (Rome and Grosseto, 1980), pp. 133–143; Antonio Gondoli and Antonio Natali (eds), *Luoghi della Toscana medicea* (Florence, 1980), p. 84; Leonardo Carandini, 'La Posta di Radicofani', *L'Universo*, Volume 44, No. 1 (1964), pp. 153–176.

74 AIL, *Archivio manoscritti*, 18, notebook VI, p. 45. About Amoretti's role as counselor to the Mines see Davide Arecco, *Mongolfiere, scienze e lumi nel tardo Settecento: Cultura accademica e conoscenze tecniche dalla vigilia della Rivoluzione francese all'età napoleonica* (Bari, 2003), pp. 163–164.

75 AIL, *Archivio manoscritti*, 18, notebook IV, pp. 297–333, in particular p. 313. About the Borromeo family and the mines in the Anzasca Valley, see Giuseppe Pipino, 'L'oro del Monte Rosa e la sua storia', *Bollettino Storico per la Provincia di Novara*, Volume 91, No. 2 (2000), pp. 321–352; Marco Del Soldato, 'Un'indagine giudiziale rivelatrice di tecniche minerarie secentesche in Valle Anzasca', *Bollettino Storico per la Provincia di Novara*, Volume 77, No. 2 (1986), pp. 111–126. On the Simonetta family, see Marco Battistoni, *L'amministrazione sabauda e i traffici commerciali nel secolo XVIII*, in Angelo Torre (ed), *Per vie di terra. Movimenti di uomini e di cose nelle società di antico regime* (Milan, 2007), pp. 109–132, in particular pp. 127–128; Edoardo Torriani (ed), 'Alcuni documenti relativi ad Emanuele Haller, in relazione al suo palazzo di Mendrisio (1794–1818)', *Bollettino Storico della Svizzera Italiana*, Volume 18 (1896), pp. 19–24, in particular p. 22.

76 AIL, *Archivio manoscritti*, 18, notebook IV, pp. 313–318.

77 Brocchi, *Giornale delle osservazioni: Volume I*, p. 235.

78 Ibid., pp. 289 (Assiut), 298–299 (Abutig), 306 (fishermen). See also Luigi Antonio Balboni, *Gl'italiani nella civiltà egiziana del secolo XIX. Storia – biografie – monografie: Volume I* (Alexandria, 1906), p. 370; Giacomo Lumbroso, *Descrittori italiani dell'Egitto e di Alessandria* (Rome, 1879), p. 96.

79 Brocchi, *Giornale delle osservazioni: Volume II*, pp. 123, 124, 148.

80 Brocchi, *Giornale delle osservazioni: Volume I*, pp. 328–331, 339–341, 346–347.

81 North, *Recollections of a Happy Life: Volume II*, pp. 220–222.

82 On Hildegonda Duckitt, see Sheila Patterson, 'Tasty Little Dishes of the Cape', in Jessica Kuper (ed), *The Anthropologist's Cookbook (Revised Edition)* (Abingdon and New York, 1997), pp. 114–121; Mary Gunn and Leslie E. Codd, *Botanical Exploration of Southern Africa* (Cape Town and Pretoria, 1981), p. 182, *ad vocem*. For the history and activities of today's Groote Post, see the dedicated website, https://www.grootepost.co.za/default.aspx?CLIENTID=3069, accessed on November 15, 2022. See also Leigh-Ann Londt, *From grape vines to yummy food, Groote Post never disappoints*, May 11, 2021, https://www.capetownetc.com/cape-town/restaurants/from-grape-vines-to-yummy-food-groote-post-never-disappoints-ready/, accessed on November 15, 2022; Georgia East, *Wines of the Cape West Coast*, January 27, 2020, https://www.timeslive.co.za/sunday-times/lifestyle/food/2020-01-27-wines-of-the-cape-west-coast/, accessed on November 15, 2022.

83 North, *Recollections of a Happy Life: Volume II*, pp. 222, 226. See also, Ena Jansen, *Like Family: Domestic Workers in South African History and Literature* (Johannesburg, 2019); Funso Afolayan, *Culture and Customs of South Africa*

(Westport CT and London, 2004), pp. 1–24; Surendra Bhana and Joy B. Brain, *Setting Down Roots: Indian Migrants in South Africa, 1860–1911* (Johannesburg, 1990); Kay Saunders (ed), *Indentured Labor in the British Empire, 1834–1920* (London and Canberra, 1983).

84 North, *Recollections of a Happy Life: Volume II*, pp. 221–224.

85 Ibid., p. 227.

86 Ibid., pp. 227–228. On Stellenbosch see Darryl G. Hart, *Reformed theology and global Christianity: the cases of South Africa and Korea*, in Michael Allen, Scott R. Swain (eds), *The Oxford Handbook of Reformed Theology* (Oxford and New York), pp. 171–186, in particular p. 174.

87 North, *Recollections of a Happy Life: Volume II*, pp. 230–232. For the historical context of this area: Nigel Worden, *Adjusting the emancipation: Freed slaves and farmers in the mid-nineteenth-century South-Western Cape*, in Wilmot G. James, Mary Simons (eds), *Class, Caste and Color: A Social and Economic History of the South African Western Cape* (New Brunswick NJ, 1992), pp. 31–39.

88 In addition to the information on today's Groote Post's offers, a more general and complex framework about rural tourism in South Africa is given in Holly Hunt and Christian M. Rogerson, *Tourism-led development and backward linkages: Evidence from the agriculture-tourism nexus in southern Africa*, in Gustav Visser, Sanette Ferreira (eds), *Tourist and Crisis* (Abingdon and New York, 2013), pp. 159–179.

89 About the articulated and problematic context of the First Boer War, its causes, dynamics, and consequences on society, institutions, and politics, the reader may find useful the bibliography in the previous notes. For a more complete picture, refer also to John Laband, *The Transvaal Rebellion: The First Boer War, 1880–1881* (Abingdon and New York, 2014).

90 Giovanni Battista Balangero, *Australia e Ceylan: Studi e ricordi di tredici anni di missione* (Torino, 1897), pp. 5–6, 181–182, 354–359.

91 Ibid., p. 287.

92 Ibid., pp. 288–294.

93 On the link between cultural tourism and religious tourism in Sri Lanka, see https://www.responsibletravel.com/holidays/sri-lanka/travel-guide/buddhism-and-cultural-tourism-in-sri-lanka, accessed on November 15, 2022; https://srilanka.travel/pilgrimage, accessed on November 15, 2022; John A. Marston, *Cambodian pilgrimage groups in India and Sri Lanka*, in Courtney Bruntz, Brooke Schedneck (eds), *Buddhist Tourism in Asia* (Honolulu, 2020), pp. 107–123; Janet Cochrane, *Responses to Continuing Crisis in Sri Lanka*, in Janet Cochrane (ed), *Asian Tourism: Growth and Change* (Oxford and Amsterdam, 2008), pp. 79–91; Malcom Crick, *Resplendent Sites, Discordant Voices: Sri Lankas and International Tourism* (Abingdon and New York, 1994).

94 Luciano Maffi, *The traveling priest: food for the spirit and food for the body*, in Jean Pierre Williot and Isabelle Blanquis (eds), *Nomadic Food: Anthropological and Historical Studies around the World* (Lanham and London, 2019), pp. 147–168.

95 ASDPv, VIII, 2, L. Marchelli, Viaggi – Taccuini.

96 Paula R. Feldman and Diana Scott-Kilvert (eds), *The Journals of Mary Shelley 1814–1844: Volume 1* (Oxford, 1987), p. 120.

97 Rachel Finnegan, *Richard Pococke's Letters from the East (1737–1740)* (Leiden, 2020); Aideen Ireland, 'Richard Pococke (1704–65), antiquarian', *Peritia: Journal of Medieval Academy of Ireland*, Volume 20 (2008), pp. 353–378.

98 George T. Stokes (ed), *Pococke's Tour in Ireland in 1752* (Dublin and London, 1891), pp. 5–4.

99 For some examples see Stokes, *Pococke's Tour in Ireland*, pp. 65–66, 80, 87–88, 144–148.

100 Brocchi, *Giornale delle osservazioni: Volume I*, pp. 233–234. For the various professionals who sailed from the port of Trieste to Egypt with him, see Brocchi, *Giornale delle osservazioni: Volume I*, pp. 1–2. See also Giampiero Berti, *Un naturalista dall'Ancien Régime alla Restaurazione: Giambattista Brocchi (1772–1826)* (Bassano del Grappa, 1988), p. 139.

101 Brocchi, *Giornale delle osservazioni: Volume I*, pp. 234, 328. About Egypt under Muhammed Ali's rule, see Khaled Fahmy, *Mehmed Ali: From Ottoman Governor to Ruler of Egypt* (London, 2009); Afaf Lutfi Al-Sayyid Marsot, *Egypt in the Reign of Muhammad Ali* (Cambridge and New York, 1984).

102 Brocchi, *Giornale delle osservazioni: Volume III*, pp. 37–38, 434. Note that Damiani was still head of the Austrian consular office in Jaffa in the early 1850s: *L'Avvisatore Mercantile: Giornale di commercio e industria*, January 17, 1852, p. 14.

103 Brocchi, *Giornale delle osservazioni: Volume III*, pp. 107–114.

104 Ibid., pp. 109, 111.

105 On this subject, see Chapter 2.

106 Charles Francis Hall, *Arctic Researches and Life among the Esquimaux: Being the Narrative of an Expedition in Search of Sir John Franklin, in the Years 1860, 1861, and 1862* (New York, 1865), pp. 194–195, 197, 200.

107 Ibid., pp. 195–199.

108 Ibid., pp. 153–157. See also Ann McElroy, *Nunavut Generations: Change and Continuity in Canadian Inuit Communities* (Long Grove, 2008), pp. 30–33.

109 Hall, *Arctic Researches*, p. 157.

110 Ibid., pp. 157–158.

111 Jon Thares Davidann, *The Limits of Westernization: American and East Asian Intellectuals Create Modernity, 1860–1960* (Abingdon and New York, 2019). For Boas's biography and scientific profile, see Han F. Vermeulen, *Before Boas: the Genesis of Ethnography and Ethnology in the German Enlightenment* (Lincoln NE, 2015), pp. 430–436.

112 North, *Recollections of a Happy Life: Volume I*, pp. 53–55.

113 North, *Recollections of a Happy Life: Volume II*, pp. 311–314. About the works for the railway see Simon Collier and William F. Sater, *A History of Chile, 1808–2002* (Cambridge and New York, 2004, 2nd edition), pp. 84–85. About the exotic urban park founded by the Cousiños and the importance of the family in the mining field, refer to William Edmundson, *A History of the British Presence in Chile: From Bloody Mary to Charles Darwin and the Decline of British Influence* (New York, 2009), pp. 156–157.

114 North, *Recollections of a Happy Life: Volume II*, pp. 327–330. On the development of the Panama Canal, a good source of information can be the classic John Saxon Mills, *The Panama Canal: A History and Description of the Enterprise* (London, 1913).

115 North, *Recollections of a Happy Life: Volume II*, pp. 314–316. About the mineral waters and the *baños* in Apoquindo see also Recaredo S. Tornedo, *Chile Ilustrado: Guía descriptiva del territorio de Chile, de las capitales de provincia, i de los puertos principales* (Valparaiso, 1872), pp. 436–438.

116 North, *Recollections of a Happy Life: Volume II*, pp. 317–318. About the role of Vicuña Mackenna in Chilean political life and his intellectual profile, see Manuel Vicuña Urrutia, *Un juez en los infiernos: Benjamín Vicuña Mackenna* (Santiago de Chile, 2009).
117 North, *Recollections of a Happy Life: Volume II*, pp. 318–320.

4 Travel Literature

4.1 Instructions for the Professional Traveler

Numerous written works existed in the eighteenth and nineteenth centuries providing tips and guidelines for those undertaking a journey for purposes of pleasure, education, scientific research, or exploration. They might be travel narratives conveying objective information through the lens of the traveler's point of view or scientific texts cataloguing the flora, fauna, or minerals of unexplored or little-known regions. There were also technical aids such as maps, astronomical tables, or bona fide guidebooks with practical information on safety and comfort. These materials contain knowledge accumulated over the centuries combined with the latest scientific and geographical knowledge. Recent works by Daniel Carey, Michael F. Robinson, Paul Smethurst, Judy A. Hayden, Alison E. Martin, and John F. Romano analyze these aspects and provide perspective on the historical sweep and dynamics of travel-related literature.[1]

Building on this historiographical and methodological framework, Chapter 4 examines the role of the type of narrative or technical material in travel experiences in natural and agricultural ecosystems in the eighteenth and nineteenth centuries. We analyze a number of case studies, focusing on the types of scientific and technical texts chosen by our travelers to fulfill their missions. We assess the value of this documentation in enabling travelers to become more attuned to these ecosystems. We also consider the creative role of these travelers in writing texts inspired by their experiences.

We have analyzed the first four travelers from other perspectives in previous chapters. They are Linnaeus, Lazzaro Spallanzani, Alexander von Humboldt, and Charles Francis Hall.

We begin with a discussion of Linnaeus and his role in developing a methodology of travel in contact with environments and societies for research purposes. Linnaeus is credited with an *Instructio Peregrinatoris* in the *Amoenitates Academicae*, a series published from 1749 to 1790 containing reprints of theses written by Linnaeus's students. Initially published as the dissertation

DOI: 10.4324/9781003230519-5

of a student named Erik Nordblad, the *Instructio* was later reissued with Linnaeus himself as author. He clearly emphasized the importance of naturalistic travel in his teachings to his students. For example, after assuming the chair of medicine at the University of Uppsala in 1741, he urged Swedish youth to explore the nature of their own kingdom before setting out to explore the rest of Europe. Assimilating his teachings in varying ways and to varying degrees, many of his disciples were great naturalist travelers to various areas of the planet, including Pehr Kalm, whose explorations in North America we have analyzed in part. However, we recall Linnaeus's activity in the naturalistic exploration of Scandinavia and Finland, especially the 1732 voyage, whose report was translated into English in 1811 by James Edward Smith, founder of the Linnaean Society in London. Specifically, the *Instructio* (in Latin) outlined a program for gathering scientific information in the lands visited, emphasizing the importance of combining the investigation of the environment with the study of the societies encountered, as well as the interaction between the latter and the former in practices such as agriculture and animal husbandry.[2]

We now turn to Lazzaro Spallanzani's trip to southern Italy in 1788 to consider the research framework he proposed in the introduction to his *Viaggi alle Due Sicilie e in alcune parti dell'Appennino*. Although it did not amount to an actual method for lithological and mineralogical field research, he did work his personal experiences into a systematic scheme.[3]

Let us also remember the context: Spallanzani distinguished himself primarily for his scientific work in biology, although he was also a professor of natural history and director of the natural history museum at the University of Pavia. His field of expertise thus also extended to the sciences of geology, lithology, and mineralogy; his 1788 trip was precisely to find mineral specimens to add to the museum's collections. He also intended to study volcanoes to increase knowledge of the origin of rocks and Earth's crust (at that time, there was a hot debate between Neptunists and Plutonists). The six volumes of the *Viaggi* (1792–1797) not only addressed Spallanzani's experiences in southern Italy but also included his explorations of the Apennines of Modena and Reggio Emilia and the Euganean Hills to study pseudovolcanic phenomena and conduct other scientific research. As he specified in the introduction to his work, after choosing the Campi Flegrei, the Aeolian Islands, and Mount Etna as the best sources of 'volcanic products' for the museum, he devoted himself to collecting and studying samples 'with the thoughtfulness, constancy, and interest which I am wont to use in my examinations of other physical objects'.[4] He thus invested serious effort to developing his interests in these branches of the natural sciences, they were not just a diversion from his main research in biology.

Spallanzani described the main steps of scientific research in his introduction, then added useful insights based on what he had personally experienced. After choosing the Neapolitan region, the Aeolian Islands, and Etna as places to collect samples in 1788, he studied in depth what travelers and naturalists 'of finer coinage' had written about it. For example, he read the report on the massive eruption of Etna in 1669 written by the scientist Giovanni Alfonso Borelli (1608–1679) at the request of the Accademia del Cimento in Florence and the Royal Society of London, of which Borelli was a member, as well as the more up-to-date *Catalogue raisonné des produits de l'Etna* (1788) by the French geologist Déodat de Dolomieu (1750–1801). With regard to Vesuvius, his readings included various reports on eighteenth-century eruptions by scientists working in Naples, such as Francesco Serao (1702–1783), Giovanni Maria Della Torre (1710–1782), Gaetano de Bottis (1721–1790), and Sir William Hamilton (1730–1803), volcanologist and envoy extraordinary of Great Britain in Naples from 1764 to 1800. There was also the updated *Saggio di lithologia vesuviana* (1790) by volcanologist Giuseppe Gioeni d'Angiò (1743–1822), dedicated to the Queen Consort of Naples Maria Carolina of Habsburg-Lorraine. This work was well known in northern Italy: in 1790–1791, three fascicles of the *Annali di chimica*—edited by Luigi Valentino Brugnatelli (1761–1818), since 1796 professor of chemistry at the University of Pavia—were almost entirely devoted to it. After a review of the scientific literature on the subject, Spallanzani nevertheless concluded that his research could add valuable insight to a budding 'volcanic science'.[5]

His field research followed a definite scheme:

> I have tried to study volcanic regions as one generally studies mountains: the structure and whole of their great masses, the lay of their different parts or of the strata which compose them, the interweave and interrelations of these strata. This is what the lithologist-researcher of mountains is most concerned to know. This has been my approach in my travels.[6]

In collecting and studying samples, Spallanzani sought to apply the most rigorous method. Where possible, he did not limit himself to collecting samples from the ground surface but dug to obtain soil from a given depth. He then compared the two types of samples, taking into account the influence on the former of weathering and 'sulfureous vapors' from fumaroles. Back in Pavia, 'in the silence of my study', he then examined them in detail with the help of magnifying glasses, describing them in detail, and performing physical and chemical tests with the help of collaborators.[7]

The instructions for scientific travel provided by Linnaeus—or at any rate developed according to the Linnaean school of thought—and those given by Spallanzani for a more circumscribed context offer two examples of the centuries-old layers of data and information on nature travel that we mentioned at the beginning of the chapter. If we consider the elements of nature-based tourism and science tourism *ante litteram* recorded on the journeys of these two scientists in Lapland and Italy as analyzed in Chapters 1 and 3, texts such as the *Istructio Peregrinatoris* or the introductory pages to Spallanzani's *Viaggi* reveal their potential as written guides for these types of tourism.

With a much broader scope than Spallanzani, naturalist Alexander von Humboldt inspired many travelers with narratives of his own wilderness travel experiences. Through his volumes published during the nineteenth century, Humboldt indeed played a very important role in creating the exotic myth of the tropical South in Europe. He narrated his travel experiences, seeking to combine personal views with scientific insight, often making use of vivid and effective iconography. Such was the case with *Voyage aux régions équinoxiales du nouveau continent*, in which Humboldt narrated his travels in Latin America from 1799 to 1804. The thirty volumes were published in Paris between 1805 and 1834 accompanied by 1,425 plates and maps. It was primarily a private undertaking and cost the large sum of 840,000 francs. Equally important was another monumental work by Humboldt, *Kosmos*, published in German in five volumes (the last posthumously) between 1845 and 1862 but soon translated into other languages to reach a broad audience. *Kosmos* comprised the series of lectures on physical geography that Humboldt gave at the University of Berlin in the late 1820s, which drew from his travel experiences and attracted a huge audience. However, as we see from our analysis of some passages in Chapter 1, even a work as early as the *Essai sur la géographie des plantes* of 1805 devoted ample space to the construction of a tropical aesthetic for European audiences.[8]

In general, Humboldt's works had a tremendous influence on Western scientists, travelers, and writers during the nineteenth century. Charles Darwin himself was greatly influenced by Humboldt's descriptions of his experience in Latin America. When Darwin set off on his round-the-world voyage on the HMS *Beagle* (1831–1836), he took with him the English translation of Humboldt's account: *Personal Narrative of Travels to the Equinoctial Regions of the New Continent*, published between 1819 and 1829. It had convinced Darwin to embark on the journey that would mark his life.[9]

Then there were printed instruments for the explorer and scientific traveler of the late nineteenth century, within a relatively advanced framework of nautical science, astronomy, geography, and cartography. When the aforementioned Charles Francis Hall explored Baffin Island from 1860 to 1862, he used texts intended to be reliable tools for explorers and geographers. In January 1861 he reported: 'My books were Bowditch's *Navigator*, Burrit's *Geography and Atlas of the Heavens*, Gillespie's *Land Surveying*, Nautical Almanac for 1861, a Bible, and *Daily Food*. My instruments were 1 telescope, 1 self-registering thermometer, 1 pocket sextant, 2 magnetic compasses, and 1 marine glass'.[10] Let us examine the titles cited by Hall.

The first work was the well-known *The American Practical Navigator*, a handbook encyclopedia of navigation written by mathematician and astronomer Nathaniel Bowditch (1773–1838) and published in its first edition in 1802 in Newburyport, Massachusetts. Its latest edition dates to 2019, updated and supplemented in 2021, for the National Geospatial-Intelligence Agency of the U.S. Department of Defense. Its longevity in print testifies to the scientific and technical validity of this encyclopedia, which 'describes in detail the principles and factors of navigation, including piloting, electronic navigation, celestial navigation, mathematics, safety, oceanography and meteorology'.[11]

The Geography of the Heavens was written by another American mathematician and astronomer, Elijah Hinsdale Burritt (1794–1838), and was first published in 1833 in Hartford, Connecticut, along with an *Atlas* illustrating it. In truth, this pair of works was intended mainly for teaching purposes, but the fact that Hall wanted it with him on Baffin Island attests to its practicality of use by an explorer. The success of this two-book set was impressive: the preface to the 1873 edition claimed that 300,000 copies had been sold.[12]

Hall also had with him *A Treatise on Land-Surveying* by William Mitchell Gillespie (1816–1868), professor of civil engineering at Union College in Schenectady, New York, and author of technical manuals. The first edition of the *Treatise* was released in 1855, although an earlier version was printed in 1851 as a textbook for Gillespie's course at Union College.[13] The work had value as a tourism instrument for those wishing to visit Europe. Indeed, the author had pursued a significant part of his engineering education at the École des Ponts et Chaussées, living for several years in France and Italy. In 1845 he published *Rome as Seen by a New-Yorker*, a veritable guidebook in which he described the Roman environment, including the local gastronomy.[14] Hall certainly had no need for a tourist guide to Rome on Baffin Island, but he kept Gillespie's *Treatise* handy should he need to calculate areas to better orient himself and give as exact a description as possible of the areas he visited.

As for the *Nautical Almanac*, it was almost certainly *The American Ephemeris and Nautical Almanac*, one of the indispensable aids to navigation, published since 1852 by a dedicated office within the U.S. Department of the Navy.[15]

The Bible, by comparison, was a source of spiritual comfort and certainly of no use as a source of information about the region Hall was exploring. However, the same cannot be said of geologist Giovanni Battista Brocchi's experience in Egypt and the Levant forty years earlier, as we shall see. As for *The Daily Food*, Hall described it as a book 'small in size' but 'which proved my solace and good companion in many a solitary and weary hour'. It was given to him by a friend on the dock in New London, Connecticut, in May

1860 as his expedition was departing for Baffin Island. It was probably a kind of almanac with edifying content.[16]

With the exception of the last two texts, most of the titles Hall had with him were authoritative technical texts for getting around in the far-from-hospitable northern environments. They were complemented by the instrumentation listed above, but Hall also availed himself of Inuit guides as discussed in Chapter 3. This organization was necessary to allow Hall to experience the ice worlds that so fascinated him, and we might consider him to be a pioneer of modern Arctic tourism.

4.2 Preparing for a Nature Trip

People preparing to travel to distant lands sought to acquire as much information as possible about their destinations, not only from specifically purposed travel instructions and technical texts but also from a wide variety of alternative sources. They thus assembled a bibliography of natural science and antiquarian texts to broaden their knowledge of the world in general and their destination in particular. They also studied travel accounts, which were not intended as instructions or technical support for others but contained anecdotal, first-person information from which further knowledge and insights could be gleaned. We find certain elements of this approach in the scientific bibliography studied by Spallanzani (Section 4.1) in preparation for his explorations in southern Italy, where he reviewed accounts of the various eruptions that occurred in the late seventeenth and eighteenth centuries, supplementing them with updated lithological descriptions of samples from Etna and Vesuvius.

In this section, we analyze works collected by travelers in relation to their explorations of natural and agricultural ecosystems. We begin with the botanist Antonio José Cavanilles, whose travel accounts of the Valencian region we have examined in Chapter 1. In particular, we present the list of volumes in his library.[17] Table 4.1 presents a selection of works from the long list that describe the flora and at times some of the fauna in territories outside Spain (in Europe or beyond). We have also included texts—present in smaller numbers—describing travel experiences.

A large number of texts described the flora of such European regions as France, Denmark, Lapland, England, Central Europe, Italy, and Spain. A similar number of scientific texts existed relating to the flora of non-European lands: the territories of the Russian Empire, Egypt, Morocco, the Dutch Cape Colony, Japan, South and Southeast Asia, and the Americas.

While not in numbers similar to those referring to the plant kingdom, Cavanilles's library also contained texts describing the fauna of certain lands, specifically the texts of Giuseppe Olivi (1769–1795) and Maria Sibylla Merian (1647–1717). The former was a naturalist from Veneto who was

Table 4.1 Selection of texts from the library of Antonio José Cavanilles. Archivo del Real Jardín Botánico de Madrid, XIII, 1, 33, 1, 'Indice de los libros de historia natural existentes en la biblioteca de Don Antonio Josef Cavanilles', November 1804.*

Author	Title	Place and date
Prospero Alpino	De Plantis Aegypti Liber	Padua 1640
Johann Ammann	Stirpium rariorum in Imperio Rutheno sponte provenientium icones et descriptiones	Saint Petersburg 1739
Jean Baptiste Christophore Fusée Aublet	Histoire des plantes de la Guiane françoise	London 1775
Pierre Bulliard	Herbier de la France ou Collection complette des plantes indigenes de ce royaume	Paris 1780–1793
Pierre Bulliard	Histoire des plantes vénéneuses de la France	Paris 1784
Johannes Burman	Rariores Plantae Africanae	Amsterdam 1738
Nikolaus Laurens Burman	Flora Indica	Leiden 1786
James Cook	Voyage dans l'hemisphere austral	Paris 1778
René Desfontaines	Flora Atlantica	Paris 1800
John Talbot Dillon	Travels through Spain	London 1782
John Ellis	Essai sur l'histoire naturelle des corallines et d'autres productions marines du même genre	The Hague 1756
Barthélemy Faujas De Saint-Fond	Mineralogie des volcans	Paris 1784
George Christian Oeder	Flora Danica	Copenhagen 1770
Samuel Gottlieb Gmelin	Flora Sibirica	Saint Petersburg 1768
Nikolaus Joseph von Jacquin	Selectarum Stirpium Americanarum Historia	Vienna no date
Nikolaus Joseph von Jacquin	Flora Austriaca	Vienna 1763
Charles-Louis L'Héritier de Brutelle	Sertum Anglicum	Paris 1788
Carl Linnaeus	Oratio de necessitate peregrinationum intra patriam	Leiden 1743
Carl Linnaeus	Flora Lapponica (ed. James Edward Smith)	London 1792
Pehr Loefling	Iter Hispanicum	Stockholm 1758
João de Loureiro	Flora Cochinchinensis	Berlin 1793
Marc Mapp	Historia Plantarum Alsaticarum	Strasbourg 1742
Maria Sibylla Merian	Dissertatio De Generatione Et Metamorphosibus Insectorum Surinamensium	Amsterdam 1779
André Michaux	Histoire des chênes de l'Amérique septentrionale	Paris 1801
Juan Ignacio Molina	Saggio sulla storia naturale del Chili	Bologna 1782
Juan Ignacio Molina	Essai sur l'histoire naturelle du Chili	Paris 1789
James Murphy	Travels in Portugal	London 1795
Domenico Nocca	Historia Ticinensis Plantae Selectae	Pavia 1800
Giuseppe Olivi	Zoologia adriatica	Bassano 1798

(Continued)

Table 4.1 (Continued)

Author	Title	Place and date
Pierre Marie François de Pagès	*Voyages autour du monde, et vers les deux pôles, par terre et par mer*	Paris 1782
Peter Simon Pallas	*Voyage… dans plusieurs provinces de l'Empire de Russie, et de l'Asie septentrionale*	Paris 1792
Peter Simon Pallas	*Flora Rossica*	Saint Petersburg 1784
Hipólito Ruiz and José Antonio Pavón	*Flora Peruviana et Chilensis sive Novorum Generum Plantarum Peruvianarum et Chilensis Descriptiones et Icones*	Madrid 1794–1798
Charles Plumier	*Description des plantes de l'Amérique*	Paris 1693
Joseph Quer	*Flora Española*	Madrid 1762
Hendrik van Rheede	*Hortus Indicus Malabaricus*	Amsterdam 1678
Albrecht Wilhelm Roth	*Tentamen Florae Germanicae*	Leipzig 1788
Georg Eberhard Rumphius	*Herbarium Amboinense*	Amsterdam 1750
Horace-Bénédict de Saussure	*Voyages dans les Alpes*	Neuchâtel 1779–1796
Peter Schousboe	*Jagttagelser over vextriger i Marokko*	Copenhagen 1800
Giovanni Antonio Scopoli	*Flora Carniolica*	Vienna 1772
Hans Sloane	*A voyage to the islands Madera, Barbados, Nieves, S. Christophers and Jamaica*	London 1707
Pierre Sonnerat	*Voyage à la Nouvelle Guinée*	Paris 1776
Pierre Sonnerat	*Voyage aux Indes Orientales et à la Chine*	Paris 1782
Olof Peter Swartz	*Dispositio Systematica Muscorum Frondosorum Sueciae*	Erlangen 1799
Olof Peter Swartz	*Flora Indiae Occidentalis*	Erlangen 1797
Carl Peter Thunberg	*Flora Japonica*	Leipzig 1784
Carl Peter Thunberg	*Prodromus Plantarum Capensium*	Uppsala 1794
Carl Peter Thunberg	*Resa uti Europa, Africa, Asia*	Uppsala 1788
Thomas Walter	*Flora Caroliniana*	London 1788
John White	*Voyage à la Nouvelle Hollande*	Paris 1798

* The place and date of edition indicated from time to time in the table are those read in the manuscript source. In the case of works consisting of several volumes, we have omitted the number of them owned by Cavanilles for convenience and because it is not necessary for our analysis.

active in the Italian and European network of natural sciences until his untimely death. His *Zoologia adriatica* provided a description of the animal species found mainly in the gulf and lagoons of Venice, with a particular focus on the seabed. His attention to Adriatic fauna was thus embedded in a more general field study of the marine ecosystem.[18]

Merian was a professional naturalist and illustrator originally from Frankfurt who later settled in Amsterdam. In 1699, she set sail with one of her daughters for the Dutch colony of Surinam on a largely self-financed mission to study and illustrate new species of arthropods. Merian had planned a five-year mission, but health problems forced her to return to Amsterdam in 1701. Nevertheless, she had obtained sufficient material to publish her *Dissertatio De Generatione Et Metamorphosibus Insectorum Surinamensium* in 1705.[19]

Merian's *Dissertatio* was clearly a very different text from Olivi's. It was, of course, much older (Cavanilles did not own the 1705 edition; see Table 4.1) and focused on a distant land outside of Europe and on a particular theme: the metamorphosis of certain insects. The scholarly literature agrees that Merian's work is not only of great current importance in gender studies on the history of science but was also a significant contribution to the subject of insect metamorphosis, then little studied.[20]

Cavanilles thus possessed a wide variety of scientific texts that provided a clear idea of numerous ecosystems. In many cases, they provided fundamental or updated knowledge of a given habitat, and the vast majority were written by authors who had direct, in-depth knowledge of them, whether European or non-European. This was the case with Merian for Surinam and, on a smaller scale, Olivi for the Adriatic coast near Venice, but we could cite numerous other cases from Table 4.1.

For example, Danish diplomat and botanist Peter Schousboe (1766–1832) led a botanical expedition to Spain and Morocco from 1791 to 1793 (from 1800 he was consul general of Denmark in Tangier, the city where he died). His descriptions of Moroccan flora were thus based on his direct travel experiences.[21] Similar dynamics characterized the work of Nikolaus Joseph von Jacquin (1727–1817), a Dutchman in the service of the Habsburgs of Austria and director of the botanical garden at the University of Vienna from 1768. Commissioned by Francis I, he traveled to areas of Latin America from 1755 to 1759 to collect plants to enrich the gardens of Schönbrunn Palace. His *Selectarum Stirpium Americanarum Historia* was published in 1763 and described the flora of some Caribbean islands, Central America, and Caribbean South America. Again, it was therefore direct experience gained on the ground by the author of the scientific text.[22]

This wealth of authentic texts on the most varied floras and faunas did not have a direct influence on the trips Cavanilles made to the Valencian region between 1791 and 1793 to study the agriculture and natural resources of that area, considering also that some of the texts considered were later than the dates of his travels. However, if the books analyzed are considered together with Cavanilles's wide network of European and non-European contacts with whom he exchanged information and plant species, we understand how his interest in useful botany proceeded hand in hand with his interest in the great variety of ecosystems existing on this planet. Therefore, the man who visited the Valencian valleys, forests, and fields and took an interest in local productions was also a scientist with a knowledge of the plant world spanning national borders (not to mention his many years of study of natural sciences in Paris).[23]

Cavanilles had also cultivated his traveler's knowledge with travel accounts on various areas around the globe by such explorers as Sloane, Sonnerat, White, and James Cook. The account of François de Pagès's travels in the late 1760s and early 1770s added experiences in environments close to the Antarctic Circle (such as the Kerguelen Archipelago in the Indian Ocean) and islands in the Arctic Circle.[24]

Also present were two authors who explored parts of Europe: Linnaeus and Saussure. Of Linnaeus's many works, Cavanilles possessed the *Flora Lapponica* and the Dutch edition of the *Oratio de necessitate peregrinationum intra patriam*: a lecture where Linnaeus urged his students to explore the local natural settings in Sweden before setting out for the rest of Europe. The text certainly had programmatic value for Cavanilles, who explored the Spanish soil of his homeland in the Valencia area. As for Saussure, in Chapter 2, we examined his role as a pioneer in mountaineering and the wealth of information that his *Voyages dans les Alpes*—in Cavanilles's library—contains regarding the Alpine ecosystem. The presence of all these works—some predating, others postdating his travels in the Valencian area in the early 1790s—is also significant of the solid culture Cavanilles had regarding travel practices in natural ecosystems.

The second traveler we consider in this section is once again Giuseppe Acerbi, a Mantuan nobleman and diplomat who visited Scandinavia and Finland as a young man in 1798–1799 and ventured as far as the North Cape into nearly untracked wilderness. In Chapters 2 and 3, we examined his view of Nordic nature and his interaction with the inhabitants of those lands; here we analyze the documentation he collected to best prepare for the journey, cross-referencing sources in his *Travels through Sweden, Finland, and Lapland, to the North Cape* (London, 1802) and in personal notes kept at the Biblioteca Comunale Teresiana in Mantua.

Acerbi became interested in a work written by Fredrik Wilhelm Radloff (1765–1838) about the Åland Islands, which lie at the entrance to the Gulf of Bothnia, which separates modern-day Sweden and Finland.[25] Of Finnish descent, Radloff was trained in natural sciences and medicine at the University of Uppsala. He later became a member of the Royal Swedish Academy of Sciences in Stockholm and taught at the Royal Academy of Åbo in Finland. His work *Beskrifning öfver Åland*, providing a topographical, naturalistic, and sociocultural description of the archipelago in question, was published in Åbo (now Turku) in 1795.[26]

During his voyage, Acerbi did not devote much time to exploring the Åland Islands. He crossed them in March 1799, proceeding on horse-drawn sleighs from the Swedish to the Finnish coast on the frozen waters of the Gulf of Bothnia. Part of the route proceeded over the islands and relied on the post stations located on them. Above all, however, Acerbi's group was concerned about continuing quickly and safely on the semi-deserted expanse of ice.[27] Radloff's book provided him with information about the climate, flora, and fauna of the archipelago, aspects in which Acerbi was particularly interested, as is clear from the notes he assembled.[28] Radloff's book thus enabled him to get an idea of the general context in which he was moving, although the season and the route prevented him from using it as a real guide. However, that and other readings, such as Linnaeus's *Flora Suecica* provided him with material to supplement the information he had gathered himself, allowing him to devote an entire chapter to the archipelago in *Travels*.[29]

Figure 4.1 Watercolor of a water bird attributed to Giuseppe Acerbi. Biblioteca Comunale Teresiana (Mantua, Italy), *Fondo Giuseppe Acerbi, Carte,* busta III, fasc. III, doc. no. 3/2.

Figure 4.2 Watercolor representing two birds attributed to Giuseppe Acerbi. Biblioteca Comunale Teresiana (Mantua, Italy), *Fondo Giuseppe Acerbi, Carte,*
busta III, fasc. III, doc. no. 3/3.

In addition to Radloff's book, among the numerous notes assembled by
Acerbi describing the naturalistic aspects of various parts of Sweden,
Finland, and more specifically Lapland, we find scattered references to other
texts that could serve both to better understand the territory he traversed
and to make the description in *Travels* more solid. These included works by
Johan Julin (1752–1820), an apothecary and naturalist operating mainly in
Oulu (Uleåborg in Swedish) on the Finnish coast of the Gulf of Bothnia,
farther north than the Åland Islands.[30] His written works, in Swedish, discussed botany, chemistry, mineralogy, meteorology, and pharmacy. He had
briefly studied natural history in Uppsala and was a member of the Royal

Swedish Academy of Sciences.[31] Acerbi personally met Julin in Oulu, the latter accompanying him on at least part of the journey. Acerbi explicitly mentioned Julin in his *Travels* in several passages concerning the climate and nature of the lands he visited.[32]

Figures like Radloff and Julin were part of a scientific community orbiting around institutions such as the University of Uppsala, the Royal Swedish Academy of Sciences, and the Academy of Åbo. These institutions had been the milieux for Linnaeus and his students, including Kalm, who played a pivotal role in defining the natural sciences and mapping of natural resources in the Kingdom of Sweden at the time. Acerbi's references thus clearly demonstrate his regard for the scientific-naturalistic disciplines developing in those lands.

We close this section with the case study of an interesting Italian botanist and traveler, Alberto Parolini (1788–1867). He came from a wealthy family with significant real estate assets and landholdings. The family also harbored a deep love for the environment and study of nature, which Parolini pursued in his university studies in Padua and Pavia. His father designed and created a garden in the northern Italian town of Bassano del Grappa, where the family originated. When his father died in 1815, Alberto Parolini was left to administer the large family estate and also the garden. He applied himself with such creativity and dedication that it was soon a destination for great European personages, including Alexander von Humboldt and the Emperor of Austria. Among the strategies for enriching the family garden, Parolini favored traveling to different countries and staying in important scientific centers to acquire new tree species. When Parolini visited Greece and Asia Minor in 1819 together with his great friend, the English botanist Philip Barker Webb (1793–1854), it was the scientific writings of the late John Sibthorp (1758–1796) on the flora of those lands that provided a solid basis for a better understanding of the Aegean environment. Sibthorp studied medicine at Oxford, Edinburgh, and Montpellier, but he was mainly interested in the plant kingdom. He decided to make a trip to Greece to identify the plants described by the first-century botanist Dioscorides. Between 1786 and 1787 he traveled through both mainland and island Greece and Cyprus, returning to Greece in 1794–1795. The second trip was fatal for him; he contracted tuberculosis and died in England the following year.[33] To prepare for his botanical travels, Sibthorp also consulted extensive documentation. For example, on his way to Greece in 1786, he stopped in Vienna to consult Dioscorides's codex, since he planned to study the Greek plants illustrated therein.[34]

In November 1818, Alberto Parolini wrote to Domenico Nocca (1758–1841), professor of botany and director of the Botanical Garden of the University of Pavia, to ask if by any chance he had a copy of the first volume of *Florae Graecae Prodromus* (London, 1806), which contained some of the material prepared for publication by John Sibthorp before his death. Parolini considered Sibthorp's research vital for his own trip. Indeed, he had ordered the volume in question from London, along with other texts, but the lot had been lost at sea. During his stay in Milan in October, Parolini had scoured the city for a copy of the text but without luck. In a last-ditch attempt, he took advantage of his acquaintance with Nocca, probably his professor in his university days, to finally obtain the scientific work of such importance for his trip.[35]

Domenico Nocca had developed an extensive network of scientific contacts over the years covering all of Europe. He studied in Pavia and Vienna and taught and did research in Mantua before returning to Pavia.[36] He was in contact with important botanists, including the aforementioned Cavanilles in Madrid, André Thouin (1747–1824) in Paris, and Carl Ludwig Willdenow (1765–1812) in Berlin. He exchanged seeds, dried specimens, information, and the most up-to-date scientific texts with these botanists. Nocca's numerous contacts in Italy were also very attuned to international progress in the natural sciences. In 1816, Filippo Re (1763–1817), professor of agricultural science at the University of Modena, who had recently acquired a number of texts from Great Britain, such as the summary catalogue of the Kew Gardens and the *Botanical Magazine*, commented on the value, both scientific and monetary, of these works with Nocca.[37] Thus, Parolini's decision to contact Nocca to obtain texts that were otherwise difficult to find was probably not dictated solely by familiarity, but also to benefit from the network Nocca had patiently woven and kept active over the years.

Note that both the two volumes of the *Prodromus* and six volumes of the *Flora Graeca* proper were organized for publication between 1806 and 1828 by the English botanist James Edward Smith. We have already had occasion to comment on Smith in this book on his role as founder and first president of the Linnaean Society in London, as well as translator into English of *Lachesis Lapponica or a Tour in Lapland*, published in London in 1811, which described Linnaeus's travel experience in Lapland. Smith had made a Grand Tour on the continent, specifically to Holland, France, and Italy between 1786 and 1787, collecting plant and insect species, visiting the most important scientific centers, and meeting fellow naturalists.[38] We therefore understand his scientific and editorial importance in boosting nineteenth-century botanical travel.

The Italian geologist Giovanni Battista Brocchi also influenced Parolini's naturalistic travels. Brocchi's influence did not come from his descriptions of his travels to Egypt, the Levant, and Sudan in the 1820s when he was mining counselor for the viceroy of Egypt, published posthumously in the 1840s, thus after Parolini and Webb's travels to Greece and Asia Minor. It should be known, however, that Brocchi was from Bassano del Grappa like Parolini, and the two were close friends. The older Brocchi influenced Parolini with his naturalistic works from his early youth, and the two made a naturalistic trip to northwestern Italy (Liguria and Piedmont) in 1813. From the data collected during this trip, Brocchi derived an important work on geology and paleontology, *Conchiologia fossile subappennina* (Milan, 1814). This was an internationally praised work, as Parolini himself informed Brocchi from the knowledge gathered during his European travels.[39]

4.3 *Ante litteram* Travel Guides

Travelers, with their various purposes, and individuals living abroad for work published their experiences in volumes (sometimes accompanied by maps and beautiful illustrative plates). These texts allowed Europeans to learn about little-known and sometimes very distant lands through the testimony of those who had been there. An excellent example of this aspect was the series of volumes of *Lettres édifiantes et curieuses*, published in Paris from 1702 to 1776. They contained the collected letters and accounts written by Jesuit missionaries on various continents to the Society's authorities and benefactors. Their publication in French, sometimes accompanied by maps and illustrations, provided European readers with greater knowledge of distant cultures and environments. The work was the initiative of Charles Le Gobien (1653–1708), who held important roles in the French management of Jesuit missions in China, and was continued by Jean-Baptiste Du Halde (1674–1743), a meticulous Jesuit scholar who studied the reports of the missionary brethren in China, although he never traveled there.[40]

Let us give an example. In 1705, the fifth volume contained a report by Sicilian Jesuit Francesco Maria Piccolo (1654–1729) about the state of missions in California. The document was translated from Spanish and had originally been presented to the Real Audiencia in Guadalajara, Mexico, in February 1702. In fact, by decree of Philip V of Spain, on whom a huge colonial empire still depended, the Audiencia was responsible for funding the missions in the area, so Piccolo produced an account of the activities receiving this subsidy. From that account an *Informe del estado de la Nueva Cristiandad de California* was published and translated into many languages.[41]

Piccolo's report, beyond the information related purely to spirituality and religion, gave succinct but compendious news also about California's geography, nature, climate, and general ecosystem. Piccolo emphasized the presence of lush pastures, pleasant valleys, springs, and numerous streams; he described the flora and fauna enthusiastically, praising their variety, the possible human uses of the various animal and plant species, from the abundant game, river fish, and sea fish, to the many fruits obtainable from the various plants.[42] We therefore understand the importance of the contribution of the *Lettres édifiantes et curieuses* in circulating information on environments outside of Europe.

Another good example from these volumes is a letter written by Father Louis-Noël de Bourzes (1673–1735), a missionary in Madurai, India, and a Tamil-language scholar.[43] In the document, he reported on a strange nocturnal twinkling he had observed at sea on his voyage from France to India in 1704, possibly the same bioluminescent dinoflagellate *Noctiluca scintillans* ([Macartney] Kofoid & Swezy, 1921) that the naturalist Pehr Kalm, a student of Linnaeus, observed in the Atlantic Ocean in 1748 on his voyage from Sweden to North America.[44] Bourzes's report on that phenomenon was rational, reviewing what he had seen and, albeit without an adequate scientific background, ventured hypotheses as to possible causes: 'It seems to me that before setting out to explain the wonders of nature, one should endeavor to learn its particulars well'.[45] However, his undeniable awe before the wonders of nature could be read between the lines. For example, he likened the luminous wake to 'a river of milk most pleasing to the eye'.[46]

Turning our attention to works published in Italy, we find the interesting case of the books on Tuscany written by Giorgio Santi (1746–1822), a professor of natural history and chemistry at the University of Pisa as well as director of its museum of natural history and botanical garden. Santi had a cosmopolitan background, with a long study stay in Paris from 1774 to 1782 and several trips along the Italian Peninsula. For his scientific work in Pisa, he made numerous studies of the environments of Tuscany, which he visited in person, producing a text on the chemical analysis of the thermal waters of the Bagni di Pisa (now San Giuliano Terme) and Asciano. He also wrote the *Viaggio al Montamiata* followed by the *Viaggio secondo* and *Viaggio terzo per le due provincie senesi*, which focused on the geological and botanical features of the Argentario and the Maremma Senese. Santi's books were an important publishing success in introducing Tuscan environments outside Italy to an audience of naturalists and potential travelers, as they already had editions in English, French, and German at the turn of the century.[47]

One of the English editions of the chemical analyses of Pisa's thermal waters was part of a small volume published in London in 1793. It was edited by physician and surgeon John Nott (1751–1825), who had traveled in Europe as a private physician and in Asia in the service of the East India Company.[48] The volume contained the 'substance' of Santi's work, accompanied by a brief guide to Pisa, with information about its history, society, places of interest, and inns collected by Nott during his visits to that city, occasions on which he had met Santi himself. This was followed by some remarks about the mineral waters of Yverdon, Switzerland, which Nott studied directly.[49]

The scholar Carlo Amoretti also made use of Santi's books when he crossed the Tuscan countryside in the fall of 1802 on his way from Milan to Rome (see Chapter 3). In particular, once he arrived in Siena in November he bought the *Viaggio al Montamiata* and the *Viaggio secondo per le due provincie senesi* from the Milanese bookseller Porri with the explicit intention of using them in the following days when he and his companions would pass through the lands studied by Santi.[50] We will recall that Amoretti was a naturalist and agriculturist and had assignments in those fields under both the Habsburgs and Napoleon, and we have reviewed his scientific interest in his surroundings on his travels, so Santi's texts served as a specialized guide for him.

The next case once again regards Giovanni Battista Brocchi. In the previous chapters, we focused on his profile and the wealth of information about the ecosystems of Egypt, Sudan, and the Levant in his *Giornale delle osservazioni* published posthumously in the 1840s. This source also allows us to gather information on the type of texts Brocchi consulted in order to better understand the ecosystems that surrounded him and contextualize the historical relationship between human society and local natural resources.

One of the sources to which Brocchi often referred was the Bible. The lands he visited included the settings for the events narrated in the Old and New Testaments. However, he treated the Bible critically, as he would any other guide to understanding the surrounding environment and society.

For example, when he was in Lebanon, Brocchi became very interested in the local flora. He studied species both from a botanical viewpoint—the symbolic cedars and other tree species—but also their practical uses as carpentry materials by local people throughout history. In studying these aspects he referred to the differences in vocabulary and information contained in the Arabic and Latin versions of the Bible, comparing them with the information he gathered during his trip.[51] Brocchi also referred to the Bible when considering certain plant species in Egypt and their uses as food. He refuted the primacy of Egyptian garlic and onions, attributing this fame to their mention in the Bible. Even in commenting on other vegetables eaten by Egyptians, such as cucurbits, he juxtaposed reference to the Bible with information he had gathered firsthand in his contact with Egyptian people.[52]

Brocchi also made numerous references to Herodotus, Aristotle, Theophrastus, Pliny, and even Virgil, in which historical and geographical information was interwoven with observations of fauna and flora. He worked for the viceroy of Egypt and recorded his travels in the 1820s, in the midst of scientific progress ushered in by the Age of Enlightenment and Napoleon's technocratic system, so his references to ancient works that were certainly not up-to-date scientific texts, especially regarding botany and zoology, might seem surprising. However, it was not unusual: Virgil was among the authorities for the European community of agriculturists.[53] And a certain classical rurality remained the model to which European academies and societies still referred in the second half of the eighteenth century and reputable agriculturists at the turn of the nineteenth.[54] Therefore, it should come as no surprise that Brocchi would refer to the previously mentioned authors in charting travel coordinates in lands still largely unfamiliar to early nine-teenth-century Europeans: Egypt, Sudan, and the Levant.

Brocchi also made abundant references to more recent naturalists and travelers, whose accounts provided a more up-to-date view of the environments in which Brocchi moved. When he traveled to Lebanon, he was also interested in the fauna. He had obtained some information from the French philosopher and orientalist Volnay, who, as mentioned in Chapter 1 and Chapter 2, visited Egypt, Palestine, and Lebanon in the 1780s and later published *Voyage en Syrie et en Égypte*. Again, Brocchi looked to this material with a critical eye. For example, he observed deer and bears that Volnay claimed were absent, as well as a widespread distribution of wolves and foxes that the Frenchman claimed were little known to the local populations. In the case of other animals, Brocchi compared the information given by Volnay with that in the Bible and that collected by himself, for example, the Arabic names of animals or his observations of living or dead specimens.[55]

Another up-to-date reference source was the 1802 *History of the British Expedition in Egypt*, authored by General Robert Thomas Wilson (1777–1849), who had participated in the defeat of Napoleon's troops in Egypt at the hands of the British in 1801.[56] He described the military maneuvers in his book with references to the historical, sociocultural, and, to some extent, environmental context of Egypt. Brocchi referred to Wilson's writing not only for information on Egypt's history but also for maps of its territory and references to the interaction between society and the environment. Despite Brocchi's general liking for Wilson, he applied the same critical gaze to the *History* as he had to other reference works, comparing it with other more or less recent texts on Egypt and with his own personal experience.[57]

Finally, it is interesting to consider some accounts of travelers with humanistic backgrounds. Who better than Mary Wollstonecraft Shelley and her *Rambles in Germany and Italy* to give us an example of bibliographic references as a guide for nature travel? First, however, it should be remembered that *Rambles* describes travels made by the author in 1840, 1842, and 1843, after her travel experiences also in other parts of Europe in the late 1810s and early 1820s. Although distant in time and along somewhat different routes, the earlier travels constituted a direct store of experience in terms of language, geography, and culture for the later travels.

The first volume of the *Rambles* begins with the words: 'I have found it a pleasant thing while travelling to have in the carriage the works of those who have passed in the same country. Sometimes they inform, sometimes they excite curiosity'. Wollstonecraft Shelley then added that her *Rambles* was intended to be 'a guide, a pioneer, or simply a fellow-traveller, for those who came after me', although she modestly pointed out that all she had to add to existing travel literature was the novelty of her own personal experience.[58]

More in the vein of the Grand Tour than other types of travel analyzed, Wollstonecraft Shelley's experience in Europe included stays and experiences in cities as well, with various kinds of cultural activities and interests in urban design and art, local society and history. However, as we highlighted in the previous chapters, she was fascinated by the varied nature of Europe. She made it clear in the introduction to *Rambles* that the natural and rural environment had played an important role in her travels and that she had tried to describe this role in her work. She cited, for example, not only the mountain passes and the 'vast extents of country' that she had passed through but also the fertility, natural beauty, and mild climate of Italy.[59]

As a means for refining her historical and sociocultural knowledge of Italy, Wollstonecraft Shelley's personal experience gave her critical insight into books written by either foreign travelers or local intellectuals. Among the authors she seemed to find most reliable were the Irish writer Lady Sydney Morgan (d. 1859) and the Neapolitan military man and historian Pietro Colletta (1775–1831). Although the two authors had completely different profiles, the output of both was characterized by direct experience of Italian society, politics, and culture, an aspect that Wollstonecraft Shelley considered trustworthy.[60]

What about the Italian environment? It is interesting to note that Lady Morgan, who visited Italy in 1819 and 1820 together with her husband, also gave information about the nature and beautiful landscapes of the peninsula in their travelogue *Italy* (1821) in addition to descriptions of the difficult political, social, and economic framework that was the focus of the work. See, for example, some of her descriptions of Lake Como, Tuscany, and Umbria, where beautiful Italian landscapes intertwine natural and man-made elements within a complicated economic and social context.[61]

Of Pietro Colletta, Wollstonecraft Shelley directly quoted *Storia del Reame di Napoli dal 1734 fino al 1825 (History of the Kingdom of Naples from 1734 until 1825)*: anything but a light read on the amenities of southern Italy. Indeed, she rightly regarded it as a significant contribution to Italian historiography. However, Colletta's *History* was not devoid of references to the environment of southern Italy and the interaction between it and human society. Here is an example referring to Charles of Bourbon, king of Naples (1734–1759) and future king of Spain (1759–1788):

> The king with the queen traveling to Castellamare by gondola and returning by land, they fell in love with the repeated views of the pleasant district of Portici, and Charles hearing that the air there was salubrious, hunting (of quail) was abundant twice a year, and the nearby sea full of fish, he commanded a villa to be built there; to one of the court who reminded him that the district was in the shadow of Vesuvius, he replied with serene spirit: 'God, Mary Immaculate, and Saint Gennaro will watch over us'.[62]

Some of Colletta's passages mentioned environmental disasters. For example, he described the hot summer of 1820 that caused fires in the forests between Terracina and Lenola, roughly halfway between Rome and Naples and not far from the Tyrrhenian coast. People planted crosses and hung votive offerings from the few surviving trees and buildings as if they were sacred sites.[63]

Italy and *History of the Realm of Naples* were thus part of the bibliographic baggage Wollstonecraft Shelley referred to in her travels to Italy in 1840, 1842, and 1843, adding her own personal experiences of some areas in the late 1810s and early 1820s. Works such as these should not be regarded as guides to be followed step by step by the writer of *Frankenstein*—the places described by those and other authors were not necessarily the same as those she visited—but they enabled her to get an accurate picture with multiple points of view not only about Italian history and society but also about its environment, its climate, and how human society interacted with it.

We close this chapter with Lady Anne Blunt (1837–1917), granddaughter of poet Lord Byron and daughter of mathematician Ada Lovelace. Blunt was a horse breeder and trader. She and her poet husband Wilfrid (1840–1922) were particularly interested in the preservation of the Arabian breed and therefore traveled extensively in the Middle East. They established a horse breeding farm in Cairo in 1882, where after their separation Lady Anne used to spend several months of the year and where she moved permanently from England in 1915 to spend the very last years of her life.[64]

In their travels to study and purchase Arabian horses, the Blunts gained considerable knowledge of Middle Eastern cultures as well as the desert environment. Arabian horses being the focus of their travels, alongside direct experience the Blunts also sought out specialized literature but did not find many satisfactory works.

For example, they read German cartographer Carsten Niebuhr (1733–1815), who was engaged for much of the 1760s on a Danish-flagged expedition that covered the Middle East and ventured as far as India. They also referred to the Swiss orientalist Johann Ludwig Burckhardt (1784–1817), who traveled the Middle East studying its intricate history and cultures from 1809 until his death from dysentery in Egypt. Although they recognized how comprehensive Niebuhr and Burckhardt were in printed accounts of their travels, the Blunts found little or erroneous information regarding Arab horse-breeding both in them and in the accounts of later travelers.[65]

Gifford Palgrave (1826–1888) provided hasty and erroneous information in this area, as the Blunts were able to verify firsthand based on their own experience. Palgrave traveled Arabia in the early 1860s as a Jesuit missionary, gathering information useful to France and its expansionist aims. The texts recounting his travels were a great success with the public, but the Blunts evidently did not consider them a satisfactory source of knowledge on

Arabian horses. Another traveler who was rather parsimonious with information in this regard was General Melchior Joseph Eugène Daumas (1803–1871). He resided for a long time in the French colony of Algeria, holding important positions. Daumas was particularly interested in horses, and a specific text of his on the subject was published in English, but the Blunts found it reductive regarding Arabian horses. They found more value in the pamphlet *Newmarket and Arabia* by Major Roger Dawson Upton (d. 1881), who had imported Arabian horses into England a few years earlier, although they had died or failed to breed.[66]

The Blunts' works chronicling their travels and experiences with Arabian horses—*Bedouin Tribes of the Euphrates* (1879) and *A Pilgrimage to Nejd* (1881)—were published by the London publisher Murray, with the contents of Anne's journals edited by her husband Wilfrid.[67] Murray also published the iconic *Handbooks for Travellers* series from 1836 through the early twentieth century, also covering the areas of the Middle East and North Africa visited both by the Blunts and by some of the authors they referred to in their work with Arabian horses. The Murray handbooks were generally written by authors with direct experience of the geographical area covered and drew on an extensive bibliography of author-travelers, including those cited by the Blunts themselves.[68]

For example, Burckhardt was widely mentioned as one of the most influential travelers to the Levant.[69] The author of the respective Murray handbook was Josias Leslie Porter (1823–1889), an Irish Presbyterian missionary with firsthand experience of that area. He had already published, also with Murray, *Five Years in Damascus*, which he repeatedly self-cited in the Levant handbook.[70] We have seen how the Blunts themselves, while finding Burckhardt's knowledge of horses lacking, considered him to be among the most comprehensive authors on travel experiences in the Middle East.

The Murray handbook devoted to Algeria, on the other hand, reads in the section on fauna: 'Regarding the horse, the reader cannot do better than study the excellent work of General Daumas, *Les Chevaux du Sahara*,' an author whom we have seen was not particularly appreciated by the Blunts for his materials on horse breeding.[71] Initially anonymous, the author of this Murray handbook was the Scotsman Robert Lambert Playfair (1828–1899), consul general in Algeria and Tunisia, who—as he stated in the preface—had traveled extensively in the region with the help of French personalities and had read up on specialized historical and archaeological journals.[72]

Another particularly long-lived series of travel guides were the German Baedeker, published from the 1850s through the 1970s. References to traveler-authors in this series sometimes reflected Murray's author choices. For example, Burckhardt was cited in the Baedeker guide devoted to Syria and Palestine but less frequently than he was in the Murray handbook written by Porter.[73] The main author in this case was the Swiss Albert Socin

(1844–1899), professor of Oriental languages at the University of Basel and later of Semitic languages at the University of Tübingen, who spent several years in the Levant and Iraq studying those languages.[74]

This production of texts explicitly devoted to travel played a central role in the evolution of the tourism phenomenon. True, the Murray and Baedeker guides—followed later by those of other publishers—gave the traveler a particular lens and a particular perspective through which to view a place. This obviously applied to both anthropogenic and natural settings and led to the preservation of prejudices.[75] However, it is also interesting to note how the authors of these important handbooks and guides expressed different evaluations of the texts left by fellow travelers in the same lands. These judgments and interpretations were shaped by their own profession and personal travel experience.

Notes

1 John F. Roman (ed), *Medieval Travel and Travellers: A Reader* (Toronto, Buffalo and London, 2020); Daniel Carey, *Advice on the Art of Travel*, and Michael F. Robinson, *Scientific Travel*, both in Nandini Das and Tim Youngs (eds), *The Cambridge History of Travel Writing* (Cambridge and New York, 2019), respectively pp. 392–407 and 488–503; Alison E. Martin, '*Fresh Fields of Exploration': Cultures of Scientific Knowledge and Ida Pfeiffer's Second Voyage round the World (1856)*, in Alison E. Martin, Lut Missinne, Beatrix van Dam (eds), *Travel Writing in Dutch and German, 1790–1830: Modernity, Regionality, Mobility* (London and New York, 2017), pp. 75–94; Paul Smethurst, *Travel Writing and the Natural World, 1748–1840* (Basingstoke and New York, 2012); Judy A. Hayden, *Travel Narratives, the New Science, and Literary Discourse, 1569–1750* (Farnham and Burlington, 2012).
2 *Instructio Peregrinatoris quam sub praesidio D. D. Car. Linnaei proposuit Ericus And. Nordblad Gevalia-Gestricius Upsaliae 1759 Maji 9*, in *Amoenitates Academicae: Volume V* (Erlangen, 1788, 2nd edition), pp. 298–313. See also: Annika Lindskog, *Constructing and Classifying 'the North': Linnaeus and Lapland*, in Cian Duffy (ed), *Romantic Norths: Anglo-Nordic Exchanges, 1770–1842* (Cham, 2017), pp. 75–99; Kenneth Nyberg, *Linnaeus's Apostles and the Globalization of Knowledge, 1729–1756*, in Patrick Manning and Daniel Rood (eds), *Global Scientific Practice in an Age of Revolutions, 1750–1850* (Pittsburgh, 2016), pp. 73–89; Han F. Vermeulen, *Before Boas: The Genesis of Ethnography and Ethnology in the German Enlightenment* (Lincoln NE, 2015), pp. 233–236; Alix Cooper, *Inventing the Indigenous: Local Knowledge and Natural History in Early Modern Europe* (Cambridge and New York, 2007), pp. 167–169; Per Sörbom, *Diderot's Russian University*, in Bo Göranzon, Magnus Florin (eds), *Skill and Education: Reflection and Experience* (London, 1992), pp. 195–206, especially p. 203; Richard Pulteney, *A General View of the Writings of Linnaeus* (London, 1805, 2nd edition), pp. 425–426.
3 Lazzaro Spallanzani, *Viaggi alle Due Sicilie e in alcune parti dell'Appennino: Volume I* (Pavia, 1792), pp. xi–lv.
4 Paolo Mazzarello, *Spallanzani, Lazzaro*, in *Dizionario Biografico degli Italiani: Volume 93* (Rome, 2018), https://www.treccani.it/enciclopedia/lazzaro-

spallanzani_%28Dizionario-Biografico%29/, accessed on December 4, 2022; Spallanzani, *Viaggi alle Due Sicilie: Volume I*, pp. xii–xiii.

5 Ibid., pp. xiii, xx–xxi, xxiii. About Borelli see Ugo Baldini, *Borelli, Giovanni Alfonso*, in *Dizionario Biografico degli Italiani: Volume 12* (Rome, 1971), https://www.treccani.it/enciclopedia/giovanni-alfonso-borelli_%28Dizionario-Biografico%29/, accessed on December 13, 2022. On Dolomieu and his contribution to geology refer to Kenneth L. Taylor, *Dolomieu, Dieudonné (called Déodat) de Gratet de*, in Charles Gillispie (ed), *Dictionary of Scientific Biography: Volume 4* (New York, 1971), pp. 149–153. About Serao refer to Marco Pantaleoni, Fabiana Console, Lorenzo Lorusso, Fabio Massimo Petti, Antonia Francesca Franchini, Alessandro Porro, Marco Romano, *Italian physicians' contribution to geosciences*, in Christopher J. Duffin, Christopher Gardner-Thorpe, Richard T. J. Moody (eds), *Geology and Medicine: Historical Connections* (London, 2017), pp. 55–75, in particular pp. 60–61. On Della Torre see Ugo Baldini, *Della Torre, Giovanni Maria*, in *Dizionario Biografico degli Italiani: Volume 37* (Rome, 1989), https://www.treccani.it/enciclopedia/della-torre-giovanni-maria_%28Dizionario-Biografico%29/, accessed on December 13, 2022. About Gaetano de Bottis see A. Nazzaro, *Il Vesuvio. Storia eruttiva e teorie vulcanologiche* (Naples, 2001), pp. 78–81. On Sir Hamilton's contribution to volcanology see David T. Moore, 'Sir William Hamilton's Volcanology and His Involvement in *Campi Phlegræi*', *Archives of Natural History*, Volume 21, No. 2 (1994), pp. 169–193. For Gioeni's contribution and his contacts with Dolomieu and Hamilton, refer to Giuseppina Buccieri, *Gioeni, Giuseppe*, in *Dizionario Biografico degli Italiani: Volume 55* (Rome, 2001); Giuseppe Gioeni, *Saggio di litologia del Vesuvio*, in *Annali di chimica ovvero Raccolta di memorie sulle scienze, arti, e manifatture ad essa relative di L. Brugnatelli*, Volume 1 (1790), pp. 243–267; Volume 2 (1791), pp. 80–160; Volume 3 (1791), pp. 160–264.

6 Spallanzani, *Viaggi alle Due Sicilie: Volume I*, pp. xiv–xv.

7 Ibid., pp. xviii–xx, xxviii, xxxiv–xxxv.

8 Enrico Banfi, Filippo Bianconi, Davide Domenici, Agnese Visconti, *Making Alexander von Humboldt's Works Known in Lombardy: The American Volumes of Giulio Ferrari's* Costume antico e modern, monographic issue of *Natura*, Volume 111, No. 2 (2021); Ronald A. Fullerton, *The Foundations of Marketing Practice: A History of Book Marketing in Germany* (Abingdon and New York, 2016), pp. 130–131; Pår Eliasson, *Humboldt, Alexander von*, in Jennifer Speake (ed), *Literature of Travel and Exploration: An Encyclopedia: Volume I* (Abingdon and New York, 2003), pp. 572–574; Nigel Leask, *Curiosity and the Aesthetics of Travel-Writing, 1770–1840* (Oxford and New York, 2002), pp. 246–256.

9 Daniela Bleichmar, *Visual Voyages: Images of Latin American Nature from Columbus to Darwin* (New Haven CT and London, 2017), pp. 157–158.

10 Charles Francis Hall, *Arctic Researches and Life among the Esquimaux: Being the Narrative of an Expedition in Search of Sir John Franklin, in the Years 1860, 1861, and 1862* (New York, 1865), p. 195.

11 See *Nautical Publications* on the website of the National Geospatial-Intelligence Agency, https://msi.nga.mil/Publications/APN, accessed on December 15, 2022. Refer also to Linda Briley-Webb, *Practical ocean navigation*, in Rosanne Welch, Peg A. Lamphier (eds), *Technical Innovation in American History: An Encyclopedia of Science and Technology: Volume 1* (Santa Barbara and Denver, 2019), pp. 137–138.

12 Elijah H. Burritt, *The Geography of the Heavens, and Class-Book of Astronomy: Accompanied by a Celestial Atlas…, Greatly Enlarged, Revised, and Illustrated by*

H. Mattison A.M. (New York, 1873), p. iii. See also: Stella Cottam and Wayne Orchiston, *Eclipses, Transits, and Comets of the Nineteenth Century: How America's Perception of the Skies Changed* (Cham, 2015), pp. 30 and 182–183; Albert J. Brooks, 'Elijah Hinsdale Burritt: The Forgotten Astronomer', *Popular Astronomy*, Volume 44 (1936), pp. 293–298.

13 William M. Gillespie, *A Treatise on Surveying Comprising the Theory and the Practice … Revised and Enlarged by Cady Staley, Ph.D., President of Case School of Applied Science: Volume I* (New York, 1896), p. iii.

14 William M. Gillespie, *Rome as Seen by a New-Yorker in 1843–1844* (New York and London, 1845). For Gillespie's training in France, see Charles D. Cashdollar, *The Transformation of Theology, 1830–1890: Positivism and Protestant Thought in Britain and America* (Princeton, 1989), p. 112.

15 Adam J. Perkins and Steven J. Dick, *The British and American Nautical Almanacs in the 19th Century*, in P. Kenneth Seidelmann, Catherine Y. Hohenkerk (eds), *The History of Celestial Navigation: Rise of the Royal Observatory and Nautical Almanacs* (Cham, 2020), pp. 157–197, in particular p. 185. There was also a British *Nautical Almanac*, published since 1767 and 'merged' with the American one in 1958: this case is analyzed in Rebekah Higgitt, *Equipping expeditionary astronomers: Nevil Maskelyne and the development of 'precision exploration'*, in Fraser MacDonald, Charles W. J. Withers (eds), *Geography, Technology and Instruments of Exploration* (London and New York, 2015), pp. 15–36.

16 Hall, *Arctic Researches*, p. xxvii.

17 For an introductory study of the formation of the Cavanilles library see Nicolás Bas Martín and María Luz López Terrada, *Una aproximación a la biblioteca del botánico valenciano Antonio José Cavanilles (1745–1804)*, in Real Sociedad Económica de Amigos del País (ed), *Antonio José Cavanilles (1745–1804): segundo centenario de la muerte de un gran botánico* (Valencia, 2004), pp. 201–285.

18 Alessandro Ottaviani, *Olivi, Giuseppe*, in *Dizionario Biografico degli Italiani: Volume 79* (Rome, 2013), https://www.treccani.it/enciclopedia/giuseppe-olivi_%28Dizionario-Biografico%29/, accessed on December 15, 2022.

19 Sarah B. Pomeroy and Jeyaraney Kathirithamby, *Maria Sibylla Merian: Artist, Scientist, Adventurer* (Los Angeles, 2018); Kay Etheridge, *The History and Influence of Maria Sibylla Merian's Bird Eating Tarantula: Circulating Images and the Production of Natural Knowledge*, in Manning and Rood (eds), *Global Scientific Practice*, pp. 54–70; Natalie Zemon Davis, *Women on the Margins: Three Seventeenth-Century Lives* (Cambridge MA, 1997), pp. 140–202.

20 Kay Etheridge, *The Flowering of Ecology: Maria Sibylla Merian's Caterpillar Book* (Leiden and Boston, 2021), pp. 3–126.

21 About Schousboe's studies on Moroccan flora see Benito Valdés, 'Early Botanical Exploration of the Maghreb', *Flora Mediterranea*, Volume 31 (2021), pp. 5–18, in particular pp. 15–17.

22 Santiago Madriñán, *Nikolaus Joseph Jacquin's American Plants Botanical Expedition to the Caribbean (1754–1759) and the Publication of the Selectarum Stirpium Americanarum Historia* (Leiden and Boston, 2013).

23 Martino Lorenzo Fagnani, 'From "Pure Botany" to "Economic Botany" – Changing Ideas by Exchanging Plants: Spain and Italy in the Late Eighteenth and the Early Nineteenth Century', *History of European Ideas*, Volume 48, No. 4 (2022), pp. 402–420.

24 For Pagès's biographical profile and an analysis of his traveling experience, see Numa Broc (ed), *Autour du monde : voyage de François de Pagès par terre et par mer, 1767–1771* (Paris, 1991).

25 BCT, *Fondo Giuseppe Acerbi, Carte*, III, 5, 2, notes in French from Radloff's book entitled 'Abrégé de la Description d'Alande par Mr. Radloff'.

26 Henry Väre, 'Fredrik Wilhelm Radloff – Demonstrator in Botany at old Åbo Akademi', *Memoranda Societatis Pro Fauna Flora Fennica*, Volume 92 (2016), pp. 92–98.

27 Giuseppe Acerbi, *Travels through Sweden, Finland, and Lapland, to the North Cape, in the Years 1798 and 1799: Volume I* (London, 1802), pp. 186–193.

28 BCT, *Fondo Giuseppe Acerbi, Carte*, III, 5, 2, pp. 8–19. See also Fredrik Wilhelm Radloff, *Beskrifning öfver Åland* (Åbo, 1795), pp. 228–244.

29 Acerbi, *Travels: Volume I*, pp. 194–201.

30 BCT, *Fondo Giuseppe Acerbi, Carte*, III, 5, 3, p. 20. Maybe this documentation was written by Julin himself and then annotated in French by Acerbi.

31 Henry Väre and Tauno Ulvinen, 'J. Julinin, K. H. Eberhardtin ja H. S. Zidbäckin julkaisemattomia kasvitietoja 1800–luvulta etenkin Oulusta ja muualta Pohjois-Suomesta', *Norrlinia*, Volume 12 (2005), pp. 1–58, in particular pp. 4–5.

32 Acerbi, *Travels: Volume I*, pp. 264–269, 354, 368–369.

33 Giuseppe Busnardo, *Alberto Parolini e la storia naturale del suo tempo*, in Alessandro Minelli (ed), *Storia naturale a Bassano (1788–1988)* (Padua, 1990), pp. 13–31; Agostino Brotto-Pastega, *Le case ed il Giardino del naturalista bassanese Alberto Parolini (1788–1867)* (Bassano del Grappa, 1988).

34 M. H. Lazarus and Heather S. Pardoe (eds), *Catalogue of Botanical Prints and Drawings at the National Museums & Galleries of Wales* (Cardiff, 2003), p. 25.

35 BUPv, *Autografi*, 4, dossier Parolini, letter by Alberto Parolini to Nocca, Bassano, November 18, 1818; *Autografi*, dossier Enrichetta Treves, letter by Enrichetta Treves to Domenico Nocca, Padua, March 1, 1811. For the botanical and cultural context of nineteenth-century Padua, see Ariane Dröscher, *Plants and Politics in Padua During the Age of Revolution, 1820–1848* (Cham, 2021).

36 Fagnani, 'From "pure botany" to "economic botany"', pp. 406–408.

37 For the correspondence from Filippo Re in question see BUPv, *Autografi*, 4, dossier Re, letters by Filippo Re to Nocca, Modena, March 17 and 27, 1816. For Nocca's correspondence and exchanges with Cavanilles, Thouin, and Willdenow, refer to the rich documentation in: BUPv, 3, dossier Cavanilles; BUPv, 4, dossiers Thouin and Willdenow; BCMHN, Ms THO 367/3; ARJB, DIV. I L.S. 28 1803, f. 118 recto – f. 140 recto; ARJB, DIV. I L.S. 32 1805, f. 10 verso – f. 13 verso.

38 Gina Douglas, *Italy in the Linnaean Collections: A Review of the Biological Specimens, Books and Manuscripts Linking Italy with Linnaeus and the Linnaean Society of London*, in Marco Beretta, Alessandro Tosi (eds), *Linnaeus in Italy: The Spread of a Revolution in Science* (Sagamore Beach, 2007), pp. 31–46.

39 Valerio Giacomini, *Brocchi, Giovanni Battista*, in *Dizionario Biografico degli Italiani: Volume 14* (Roma, 1972), https://www.treccani.it/enciclopedia/giovanni-battista-brocchi_%28Dizionario-Biografico%29/, accessed on December 4, 2022.

40 Marcia Reed, *A perfume is best from afar: publishing China for Europe*, in Marcia Reed and Paola Demattè (eds), *China on Paper: European and Chinese Works from the Late Sixteenth to the Early Nineteenth Century* (Los Angeles, 2007), pp. 9–27; Ouyang Zhesheng, 'The "Beijing Experience" of Eighteenth-Century French Jesuits: A Discussion Centered on *Lettres édifiantes et curieuses écrites des missions étrangères*', *Chinese Studies in History*, Volume 46, No. 2 (2012), pp. 35–57; Adrien Paschoud, *Le Monde amérindien au miroir des Lettres édifiantes et curieuses* (Oxford, 2000); David Morgan, 'Sources of Enlightenment: The Idealizing of China in the Jesuits' *Lettres édifiantes* and Voltaire's *Siècle de Louis XIV*', *Romance Notes*, Volume 37, No. 3 (1997), pp. 263–272.

41 Emanuele Colombo, *Piccolo, Francesco Maria*, in *Dizionario Biografico degli Italiani: Volume 83* (Rome, 2015).

42 Francesco Maria Piccolo, *Mémoire touchant l'éstat des missions nouvellement éstablies dans la Californie par les pères de la Compagnie de Jésus*, in *Lettres édifiantes et curieuses: Volume V* (Paris, 1705), pp. 248–287, in particular pp. 265–172.

43 Manonmani Restif-Filliozat, 'The Jesuit Contribution to the Geographical Knowledge of India in the Eighteenth Century', *Journal of Jesuit Studies*, Volume 6, No. 1 (2019), pp. 71–84; Gregory James, *Tamil Lexicography* (Tübingen, 1991), pp. 70–73.

44 In order to compare the two experiences, see: Louis-Nöel Bourzes, *Lettre au père Éstienne Souciet*, in *Lettres édifiantes et curieuses: Volume IX* (Paris, 1711), pp. 359–375; Adolph B. Benson (ed), *Peter Kalm's Travels in North America: The English Version of 1770: Volume 1* (New York, 1966), p. 14 and note 2.

45 Bourzes, *Lettre au père Éstienne Souciet*, p. 361.

46 Ibid., p. 364.

47 Renato Pasta, *Santi, Giorgio*, in *Dizionario Biografico degli Italiani: Volume 90* (Rome, 2017); Umberto Bindi, *Giorgio Santi, scienziato pientino del Settecento. Biografia e scritti inediti* (Pienza, 2014).

48 Frederick Bussby, 'John Nott, M.D. 1751–1825: The Oriental Travels of an Eighteenth-Century Physician', *Medical History*, Volume 18, No. 3 (July 1974), pp. 294–298.

49 John Nott and Giorgio Santi, *A Chemical Dissertation on the Thermal Waters of Pisa, and on the Neighbouring Acidolous Spring of Asciano with a Historical Sketch of Pisa, and a Meteorological Account of Its Weather to Which Are Added Analytical Papers Respecting the Sulphureous Waters of Yverdun* (London, 1793).

50 AIL, Archivio manoscritti, 18, notebook VI, pp. 31, 35, 41 and 45.

51 Giovanni Battista Brocchi, *Giornale delle osservazioni fatte ne' viaggi in Egitto, nella Siria e nella Nubia: Volume III* (Bassano del Grappa, 1842), pp. 365–373.

52 Giovanni Battista Brocchi, *Giornale delle osservazioni fatte ne' viaggi in Egitto, nella Siria e nella Nubia: Volume IV* (Bassano del Grappa, 1843), pp. 25–26.

53 For instance, see Mauro Ambrosoli, *The Wild and the Sown: Botany and Agriculture in Western Europe, 1350–1850* (Cambridge and New York, 1997), p. 351.

54 Martino Lorenzo Fagnani, 'Travels and Representations at the Core of Western Agricultural Science: Discovering Rural Societies in Spain, Italy and Lebanon, Late Eighteenth and Early Nineteenth Centuries', *Continuity and Change*, Volume 37, No. 3 (December 2022), pp. 313–334.

55 Brocchi, *Giornale delle osservazioni: Volume III*, pp. 373–376.

56 Michael Glover, *A Very Slippery Fellow – The Life of Sir Robert Wilson 1777–1849* (Oxford, 1978).

57 Giovanni Battista Brocchi, *Giornale delle osservazioni fatte ne' viaggi in Egitto, nella Siria e nella Nubia: Volume I* (Bassano del Grappa, 1841), pp. 64–68; Volume IV, pp. 129–130.

58 Mary Wollstonecraft Shelley, *Rambles in Germany and Italy in 1840, 1842, and 1843: Volume I* (London, 1844), pp. vii-viii.

59 Ibid., pp. vii, xiii, xvi.

60 Ibid., p. x; Volume II, pp. 205–209. See also: Emily W. Sunstein, *Mary Shelley: Romance and Reality* (Baltimore, 1989), pp. 137, 205, 337; Alfonso Scirocco, *Colletta, Pietro*, in *Dizionario Biografico degli Italiani: Volume 27* (Rome, 1982), https://www.treccani.it/enciclopedia/pietro-colletta_%28Dizionario-Biografico %29/, accessed on December 16, 2022.

61 Sydney Morgan, *Italy: Volume I* (Paris, 1821), pp. 267–284; Volume II, pp. 267–268, 279. See also Julie Donovan, *Sydney Owenson, Lady Morgan and the Politics*

of Style (Bethesda, Dublin and Palo Alto, 2009); Donatella Abbate Badin, *Lady Morgan's* Italy: *Anglo-Irish Sensibility and Italian Realities* (Bethesda, 2007); Anne E. O'Brien, *Lady Morgan's travel writing on Italy: a novel approach*, in Jane Conroy (ed), *Cross-Cultural Travel: Papers from the Royal Irish Academy Symposium on Literature and Travel* (New York, 2003), pp. 179–185.

62 Pietro Colletta, *Storia del Reame di Napoli dal 1734 sino al 1825: Volume I* (Capolago CH, 1834), p. 109.

63 Ibid., p. 373. See also *Dell'assoluta influenza delle foreste sulla temperatura atmosferica*, in Adolfo di Bérenger (ed), *Giornale di economia forestale ossia raccolta di memorie lette nel R. Istituto Forestale di Vallombrosa: Volume I* (Florence, 1871–1872), p. 60.

64 Lisa McCracken Lacy, *Lady Anne Blunt in the Middle East: Politics, Travel and the Idea of Empire* (London and New York, 2018); Michael D. Berdine, *The Accidental Tourist, Wilfrid Scawen Blunt, and the British Invasion of Egypt in 1882* (Abingdon and New York, 2005).

65 Anne Blunt, *Bedouin Tribes of the Euphrates: Volume II*, Wilfrid Scawen Blunt (ed) (London, 1879), p. 243. For the different contexts of Niebuhr's and Burckhardt's experiences, among the most recent studies, see Lawrence J. Baak, *Undying Curiosity: Carsten Niebuhr and the Royal Danish Expedition to Arabia (1761–1767)* (Stuttgart, 2014); Jacob Rama Berman, *American Arabesque: Arabs, Islam, and the Nineteenth-Century Imaginary* (New York and London, 2012), pp. 78–79, 108.

66 Blunt, *Bedouin Tribes: Volume II*, pp. 244–245. See also: McCracken Lacy, *Lady Anne Blunt in the Middle East*, p. 60; Daniel Foliard, *Dislocating the Orient: British Maps and the Making of the Middle East, 1854–1821* (Chicago and London, 2017), pp. 34–35; Leon B. Blair, 'The Origin and Development of the Arabian Horse', The Southwestern Historical Quarterly, Volume 68, No. 3 (January 1965), pp. 303–316; William Henry Upton, *Upton Family Records: Being Genealogical Collections for an Upton Family History* (London, 1893), pp. 98, 136.

67 McCracken Lacy, *Lady Anne Blunt in the Middle East*, p. 240.

68 About the influence of Murray's *Handbooks* on modern tourist practices refer to Innes M. Keighren, Charles W. J. Withers, Bill Bell, *Travels into Print: Exploration, Writing, and Publishing with John Murray, 1773–1859* (Chicago and London, 2013); Gráinne Goodwin and Gordon Johnston, 'Guidebook Publishing in the Nineteenth Century: John Murray's *Handbooks for Travellers*', *Studies in Travel Writing*, Volume 17, No. 1 (2013), pp. 43–61.

69 *A Handbook for Travellers in Syria and Palestine; Including an Account of the Geography, History, Antiquities, and Inhabitants of These Countries, the Peninsula of Sinai, Edom, and the Syrian Desert; with Detailed Descriptions of Jerusalem, Petra, Damascus, and Palmyra: Volume I* (London, Murray, 1858), pp. xxi, lxii.

70 For a brief biography of Porter, see Edwin James Aiken, *Revd. Prof. Josias Leslie Porter 1823–1889*, in Hayden Lorimer, Charles W.J. Withers (eds), *Geographers: Biobibliographical Studies: Volume 26* (London and New York, 2007), pp. 67–78.

71 *A Handbook for Travellers in Algeria. With Map and Plan* (London, 1874), p. 39.

72 Nabila Oulebsir, *From ruins to heritage: the past perfect and the idealized antiquity in North Africa*, in Gábor Klaniczay, Michael Werner, Ottó Gecser (eds), *Multiple Antiquities – Multiple Modernities: Ancient Histories in Nineteenth Century European Cultures* (Frankfurt and New York, 2011), pp. 335–364, especially pp. 344–345. See also the above-mentioned *A Handbook for Travellers in Algeria*, 'preface', no page numbers.

73 For instance, see *Palestine and Syria. Handbook for Travellers* (Leipzig and London, 1876), pp. 124, 291, 337.

74 Ibid., p. v. For further information about Socin's travels and academic activity refer to Andreas Bigger, *Socin, Albert*, in *Historisches Lexikon der Schweiz (HLS)* (Basel, 2011), https://hls-dhs-dss.ch/de/articles/041667/2011-11-21/, accessed on December 16, 2022.

75 Eric G. E. Zuelow, *A History of Modern Tourism* (London and New York, 2016), pp. 75–79; Lori Brister, *The precise and the subjective: the guidebook industry and women's travel writing in late nineteenth-century Europe and North Africa*, in Clare Broome Saunders (ed), *Women, Travel Writing, and Truth* (New York and Abingdon, 2014), pp. 61–76; David M. Bruce, *The nineteenth-century 'golden age' of cultural tourism: how the beaten track of the intellectuals became the modern tourism trail*, in Melanie Smith, Greg Richards (eds), *The Routledge Handbook of Cultural Tourism* (Abingdon and New York, 2013), pp. 11–18.

Conclusion

In this book we have analyzed early forms of tourism in natural and agricultural environments in the eighteenth and nineteenth centuries. The trips we have studied were typically undertaken for purposes of scientific research, work, economic affairs, or missionary activity. Today we would call this 'incidental tourism' in that the purpose was something other than a quest for relaxation or recreation. This category now applies to all those who travel for business, work, conferences, or health. In greater detail, we may also include medical or health tourism, congress tourism, business and service tourism, tourism for family-related reasons, scientific and research tourism, and study tourism. In some cases, the forms of nature tourism and rural tourism we have examined were intertwined with educational travel or the Grand Tour proper. In these latter contexts, we note that the literature has tended to focus on the role of urban centers, artistic attractions, and interchanges among intellectuals and dedicated much less attention to contact with pristine or anthropic natural settings.[1]

We have chosen the term *tourism* to encompass all the experiences analyzed in this book, but we admit that it was certainly a very different kind of tourism from today's prevailing concept. Indeed, scholars of tourism history point out that tourism practices have a very long history and a complex development. For instance, in Western culture, these practices proceeded from the *villeggiatura* of the ancient Romans to the Grand Tour in the Early Modern era, from forms of modern tourism starting in the mid-nineteenth century to mass tourism starting in the mid-twentieth century. Tourism practices, moreover, are constantly evolving, with an ever-growing quest for 'experiential' elements that make these experiences unique and memorable.[2]

In this context, it was during the course of the eighteenth and nineteenth centuries that many characteristic aspects of what are now considered nature tourism and rural tourism emerged. These two categories, each having their own distinct features, share a *raison d'être* in the desire for immersion in a given environment and its flora and fauna.

DOI: 10.4324/9781003230519-6

In considering the contact between travelers and nature, we have examined a range of different reactions to it. Some travelers are enraptured by the wilderness and find travel in natural environments to be a unique and enriching experience. We find this attitude expressed more or less explicitly in travelers such as Alexander von Humboldt, Jean Baptiste Bory de Saint-Vincent, Giovanni Battista Brocchi, and Charles Francis Hall, who experienced immersion in nature as a unique opportunity for personal growth: even in harsh environments, such as Brocchi's experience in the Egyptian desert in the early 1820s or Hall's encounter with the frozen lands near the Arctic Circle starting in 1860.

In contrast, there are travelers who did not find their wilderness experience entirely enjoyable or satisfying. The Count de Volney was rather negative about his experience in the Egyptian desert or along the Nile River, using a very different tone from what we find in Brocchi's accounts. Elizabeth Bruce Elton Smith experienced the Indian jungle in the late 1820s and early 1830s as a dangerous and hostile environment, although it did fascinate her to some extent as a place of refuge from the crowded, bustling, urbanized world. We also saw how Marianne North, while having a strong spirit of adaptation and considerable physical endurance, tended to prefer a balance between the beauty of wild nature and the comforts of the 'civilized world' of the late nineteenth century in North America as in Chile, in South Africa, and in the Seychelles.

However, regardless of whether a given traveler is enthusiastic about unspoiled nature, finds it more a source of inconvenience than anything else, or prefers to complement wild beauty with the comforts of civilization, the cases we have examined all share a strong involvement of the senses. Many are the references to colors and lights. Think of Alexis-Marie de Rochon and the colors of the plant species of Madagascar, Pehr Kalm and the luminescence of *Noctiluca scintillans* on the waters of the Atlantic, George Finlayson surrounded by the multifarious fauna of the Thai islands, Charles Francis Hall and the awe he felt before the northern lights, and Giovanni Battista Balangero fascinated by the beauty and elegance of the Sinhalese plants.

In addition to visual stimuli, many of the case studies contain an emphasis on tasting food and placing it in its appropriate sociocultural and environmental context. Mediterranean Europe offers a variety of examples, such as Carlo Amoretti sampling rustic cuisine as he traveled through the countryside of central Italy in the early nineteenth century or Luigi Marchelli in the latter half of the nineteenth century, whose passion for gardens and urban parks was complemented with that for traditional cuisine. We can also suppose that Antonio José Cavanilles had a good picture of the horticultural products and honey varieties of eastern Spain, as did Alberto Fortis of the fish dishes of Dalmatia. Regarding these two scholars, we find their comments on food in their formal printed works, thus conveyed with a more detached, objective attitude than what we have found in the more effusive personal notebooks of Amoretti and Marchelli.

However, food and food-related practices are also evidenced in travels at other latitudes. We commented on how Linnaeus, when he explored Lapland to study its flora and fauna in 1732, also took note of local food resources—from fish to reindeer and grouse meat, from dairy products to vegetables—and how they were stored and prepared for eating. Much farther south, Giovanni Battista Brocchi sampled and described foods and drinks from Sennar in the 1820s, while Marianne North enjoyed foods and wine from the Groote Post in South Africa in the early 1880s.

It is also necessary to reiterate that the embryonic forms of nature and rural tourism examined in the case studies in this book, from pescatourism to scientific tourism, were interwoven with numerous aspects of experiential tourism revolving around the human–nature relationship. However, the type of modern-day tourism to which we have most often referred throughout the book is extreme tourism in natural settings, inextricably intertwining sporting activities and risk. These elements of extreme tourism were in most cases related to the need to proceed by the only viable route, no matter how risky it was. The two most significant cases are the upriver canoe journeys by Giuseppe Acerbi and Mary Kingsley: the former ascended the Tornio River in Lapland, at times on foot along the banks in the more treacherous stretches, the latter the Ogooué rapids in Gabon. Both inevitably relied on local people who knew how to handle the watercraft and safely negotiate the rapids. In her account, Kingsley not only conveys her spirit of adventure but also her fascination with the forests of Gabon and the waters of the Ogooué as seen from the canoe.

Also within the framework of extreme tourism *ante litteram* is Darwin's experience in a Tahitian valley in 1835, proceeding along sheer rock walls past waterfalls plunging from above into the gorges below. Here too, local guides and the use of ropes were key elements in ensuring relative safety. As in Kingsley's river experience, admiration for surroundings seen through the lens of risk emerges in Darwin's account.

The proto-tourist experience in mountain environments is another element in several of the cases analyzed in our book. But not all ventures into these settings may be considered an early form of extreme tourism. In some cases—Lazzaro Spallanzani's ascent of Etna among rugged lava formations and fissures still containing still-glowing lava or Darwin's ascent of a steep Chilean cone with trees as climbing aids—a certain taste for the extreme is present. However, in the many experiences in the Alps discussed in Chapter 2—from Horace Bénédict de Saussure in the late eighteenth century to the Shelleys in the early nineteenth century to the various accounts by British mountaineers during the rest of the century—we notice a greater inclination to visit mountain ecosystems in order to study their flora, fauna, and minerals and enjoy the stunning views of the peaks. But these journeys were undertaken slowly and patiently with local guides.

We also emphasize two other important elements emerging from our case studies: logistics challenges and gender challenges. The first regard difficulties in organizing travel, which were particularly acute in journeys far afield in predominantly wilderness contexts. In the early eighteenth century, Cornelis De Bruijn found obstacles in traveling from Texel to Archangel around Norway and into Russia because of international tensions due to the Great Northern War. His journey was a sort of unconventional Grand Tour, driven by a desire for personal growth, as well as an interest in art and a desire to learn about distant lands.

Longer sea voyages had their own additional complications. One example is the ship *Duff* carrying evangelical missionaries from the London Missionary Society, departing from the English capital in 1796 bound for the Pacific islands. The documentation we have examined highlights both the uncertainty of such a long voyage and the dangers encountered even before leaving European waters due to tensions between revolutionary France and the powers in the first anti-French coalition. The missionaries' voyage was characterized by an unpleasant relationship with the oceanic environment, which was faced at great risk and with great effort. It clearly evidences the difficulties of ocean travel. So when instead we read the account of the Jesuit missionary Louis-Noël de Bourzes—who set sail from France in 1704, bound for India—about the beauty of the ocean immersed in the night, we must nevertheless realize that his voyage was probably as fraught with risk as the *Duff*'s.

The second element relates to the gender of our travelers. We may reasonably assume that the material difficulties of travel were a greater challenge to women than to men.[3] However, European women travelers were able to carve out their own identity and autonomy in their contact with exotic nature and in the narration of their experiences. Two examples are Elizabeth Bruce Elton Smith and Maria Graham. The former followed her military husband to India and published two books referring to her experience there, where the relationship with the wilderness played an important role. Graham, by comparison, traveled the world initially accompanying her father and then her first husband, both of whom were naval officers. When her first husband died, she continued to live in Latin America; as a writer, she gave great importance to the relationship between human society and the environment in those regions.

Other women travelers discussed in this book were strongly characterized by autonomy, starting with their initial decision to travel for the purpose of studying environments far from Europe and how human societies interacted with them. Two examples here are Marianne North, in the field of scientific illustration, and Mary Kingsley in anthropology. While operating in very different geographical areas—Kingsley's journeys being not only more circumscribed but also more immersive than those of her counterpart—each of them expressed their strong personalities both in the decisions they made in life and in the accounts of their travels.

This book aims to foster debate among tourism historians on the deep origins of travel in natural and agricultural environments. The experiences analyzed in this book have a Western historical and sociocultural background, all figures discussed herein hailing from Europe or North America (all Caucasian and without Amerind heritage, such as Charles Francis Hall). An analysis of perspectives beyond this context would be a valuable addition to the debate on the historical evolution of nature tourism and rural tourism. For example, recent studies on the development of nature and rural tourism in China from its origin to the contemporary period reveal many implications for understanding current and future trends in tourism development.[4]

Changing values and tastes have contributed particularly to the appreciation of the landscape as a new form of leisure since ancient times. Consider the change in attitudes toward mountains in Europe during the nineteenth century, when alpine and wilderness activities emerged as new forms of tourism. Sightseeing, climbing, and hiking in the mountains, as well as the popularity of mountain sports, can be seen as the results of the changed attitude toward the natural world. Tourism is a way of approaching nature, not only in our contemporary worldview but also through history. As discussed in this book, recreation and personal growth in the context of natural and agricultural ecosystems has a long tradition. Indeed, our analysis of the development of this kind of tourism in the eighteenth and nineteenth centuries has implications for understanding current and future trends in tourism development. Visits to natural and agricultural ecosystems have since developed characteristics that have particularly influenced tourism in the twentieth and twenty-first centuries. Nature tourism and rural tourism attract people seeking immersion in a rich natural, cultural, and historical experience. Our case studies demonstrate the plurality of motivations and interests that characterized the pioneers of this type of tourism, including their emotional experience.

A sentence from Mary Wollstonecraft Shelley's *Rambles* (1844) clearly expresses the multifaceted and sometimes conflicting fascination nature exerts on the travelers we have analyzed in this book. Comparing the peaks and glaciers of the Alps with the gentleness of the Mediterranean landscape in southern Italy, Wollstonecraft Shelley commented: 'There, nature is sublime, but she shows the power and the will to harm; here she is gracious as well as glorious; she is our friend, or rather our exalted and munificent queen and benefactress'.[5]

Notes

1 For instance, Rosemary Sweet, *Cities and the Grand Tour: The British in Italy, c. 1690–1820* (Cambridge and London, 2012).
2 On the evolution of the tourism phenomenon, see among the most up-to-date studies: Eric G. E. Zuelow and Kevin J. James (eds), *The Oxford Handbook of the History of Tourism and Travel* (Oxford, 2022); Prokopis A. Christou, *The History and Evolution of Tourism* (Wallingford and Boston, 2022); Eric G. E. Zuelow, *A History of Modern Tourism* (London and New York, 2016); Melanie Smith and Greg Richards (eds), *The Routledge Handbook of Cultural Tourism* (Abingdon and New York, 2013).
3 Emma Gleadhill, *Taking Travel Home: The Souvenir Culture of British Women Tourists, 1750–1830* (Manchester, 2022); Patricia Akhimie and Bernadette Andrea (eds), *Travel and Travail: Early Modern Women, English Drama, and the Wider World* (Lincoln and London, 2019); Emma Robinson-Tomsett, *Women, Travel and Identity: Journeys by Rail and Sea, 1870–1940* (Manchester, 2013).
4 Chunyan Zhang, David W. Knight, Yajuan Li, Yi Zhou, Meng Zhou, Minggui Zi, 'Rural Tourism and Evolving Identities of Chinese Communities in Forested Areas', *Journal of Sustainable Tourism* (2022), https://doi.org/10.1080/09669582.2022.2155829, accessed on December 22, 2022; Libo Yan, 'Origins of Nature Tourism in Imperial China', *Journal of Tourism Futures*, Volume 4, No. 3 (2018), pp. 265–274.
5 Mary Wollstonecraft Shelley, *Rambles in Germany and Italy in 1840, 1842, and 1843: Volume II* (London, 1844), p. 291.

Bibliography

Primary sources

Archives

Archivio Doria, Dipartimento di Economia dell'Università. Genoa
Archivio dell'Istituto Lombardo Accademia di Scienze e Lettere. Milan
Archivo del Real Jardín Botánico. Madrid
Archivio Storico Diocesano. Pavia
Bibliothèque Centrale du Muséum National d'Histoire Naturelle. Paris
Biblioteca Comunale Teresiana. Mantua
Biblioteca Nazionale Braidense. Milan
Biblioteca Universitaria. Pavia.

Printed sources

A Handbook for Travellers in Algeria. With Map and Plan (London, 1874).
A Handbook for Travellers in Syria and Palestine; Including an Account of the Geography, History, Antiquities, and Inhabitants of These Countries, the Peninsula of Sinai, Edom, and the Syrian Desert; with Detailed Descriptions of Jerusalem, Petra, Damascus, and Palmyra: Volume I (London, Murray, 1858).
Acerbi, Giuseppe, *Il viaggio in Lapponia (1799)*, De Anna, Luigi G., Lindgren, Lauri (eds) (Turku, 2009. 2nd edition).
Acerbi, Giuseppe, *Il viaggio in Svezia e in Norvegia (1799–1800)*, Lindgren, Lauri, De Anna, Luigi, G. (eds) (Turku, 2000).
Acerbi, Giuseppe, *Travels through Sweden, Finland, and Lapland, to the North Cape, in the Years 1798 and 1799: Volume I* (London, 1802).
Acerbi, Giuseppe, *Viaggio in Svezia e in Finlandia (1798–1799)*, Lindgren, Lauri (ed) (Turku, 2005).
Acerbi, Giuseppe, *Vues de la Suede, de la Finlande, et de la Lapponie, depuis le detroit du Sund jusq'au Cap-Nord* (Paris, 1803).
Amoretti, Carlo, *Viaggio da Milano a Nizza… ed altro da Berlino a Nizza e ritorno da Nizza a Berlino di Giangiorgio Sulzer fatto negli anni 1775 e 1776* (Milan, 1819).
Amoretti, Carlo, *Viaggio da Milano ai tre Laghi Maggiore, di Lugano e di Como, e ne' monti che li circondano* (Milan, 1794).

Balangero, Giovanni Battista, *Australia e Ceylan: Studi e ricordi di tredici anni di missione* (Turin, 1897).

Blunt, Anne, *Bedouin Tribes of the Euphrates: Volume II*, Wilfrid Scawen Blunt (ed) (London, 1879).

Bory, de Saint-Vincent, and Jean, Baptiste, *Voyage to, and Travels through the Four Principal Islands of the African Seas: Performed by Order of the French Government, during the Years 1801 and 1802: Volume II* (London, 1805).

Brocchi, Giovanni Battista, *Giornale delle osservazioni fatte ne' viaggi in Egitto, nella Siria e nella Nubia: Volumes I, II, III, and IV* (Bassano del Grappa, 1841–1843).

Browne, William George, *Travels in Africa, Egypt, and Syria, from the Year 1792 to 1798* (London, 1799).

Burritt, Elijah H., *The Geography of the Heavens, and Class-Book of Astronomy: Accompanied by a Celestial Atlas…, Greatly Enlarged, Revised, and Illustrated by H. Mattison A.M.* (New York, 1873).

Cavanilles, Antonio José, *Observaciones sobre la historia natural, geographía, agricultura, población y frutos del Reyno de Valencia: Volume I* (Madrid, 1795).

Cavanilles, Antonio José, *Observaciones sobre la historia natural, geographía, agricultura, población y frutos del Reyno de Valencia: Volume II* (Madrid, 1797).

Colletta, Pietro, *Storia del Reame di Napoli dal 1734 sino al 1825: Volume I* (Capolago CH, 1834).

Conway, William Martin, *The Alps from End to End* (Westminster, 1895).

Damoiseau, Louis, *Voyage en Syrie et dans le désert* (Paris, 1833).

Darwin, Charles, *Journal and remarks, 1832–1836*, in *Narrative of the Surveying Voyages of His Majesty's Ships Adventure and Beagle, between the Years 1826 and 1836, Describing their Examination of the Southern Shores of South America and the Beagle's Circumnavigation of the Globe: Volume III* (London, 1839).

David, Andrew, Fernandez-Armesto, Felipe, Novi, Carlos, and Williams, Glyndwr (eds), *The Malaspina Expedition 1789–1794: the Journal of the Voyage by Alessandro Malaspina: Volume 1* (London and Madrid, 2001).

De Bruijn, Cornelis, *Travels into Muscovy, Persia, and Part of the East-Indies: Volume I* (London, 1737).

de La Rochefoucauld-Liancourt, François, *Travels through the United States of North America and the Country of the Iroquois and Upper Canada in the Years 1795, 1796, and 1797; with an Authentic Account of Lower Canada: Volume I* (London, 1799).

de La Rochefoucauld-Liancourt, François, *Travels through the United States of North America and the Country of the Iroquois and Upper Canada in the Years 1795, 1796, and 1797; with an Authentic Account of Lower Canada: Volume III* (London, 1800, 2nd edition).

de La Rochefoucauld-Liancourt, François, *Voyage dans les États-Unis d'Amérique fait en 1795, 1796 et 1797: Volume I* (Paris, 1798–1799).

De Riseis, Giovanni, *Il Giappone moderno* (Milan, 1896).

de Rochon, Alexis-Marie, *A Voyage to Madagascar and the East Indies Translated from the French* (London, 1792).

de Saussure, Horace-Bénédict, *Voyages dans les Alpes, précédés d'un essai sur l'histoire naturelle des environs de Genève: Volume IV* (Geneva, 1786).

Dickens, Charles, *Pictures from Italy* (London, 1846).

Fortis, Alberto, *Travels into Dalmatia [...] Observations on the Island of Cherso and Osero. Translated from the Italian under the Author's Inspection* (London, 1778).

Giglioli, Enrico Hillyer, *Viaggio intorno al globo della r. pirocorvetta italiana Magenta negli anni 1865-66-67-68* (Milan, 1875).

Gillespie, William M., *Rome as Seen by a New-Yorker in 1843–4* (New York and London, 1845).

Gioeni, Giuseppe, 'Saggio di litologia del Vesuvio', *Annali di chimica ovvero Raccolta di memorie sulle scienze, arti, e manifatture ad essa relative di L. Brugnatelli*, Volume 1 (1790), pp. 243–267.

Gioeni, Giuseppe, 'Saggio di litologia del Vesuvio', *Annali di chimica ovvero Raccolta di memorie sulle scienze, arti, e manifatture ad essa relative di L. Brugnatelli*, Volume 2 (1791a), pp. 80–160.

Gioeni, Giuseppe, 'Saggio di litologia del Vesuvio', *Annali di chimica ovvero Raccolta di memorie sulle scienze, arti, e manifatture ad essa relative di L. Brugnatelli*, Volume 3 (1791b), pp. 160–264.

Goethe's Travels in Italy: toghether with His Second Residence in Rome and Fragments on Italy (London, 1885).

Gozzano, Guido, *Verso la cuna del mondo: Lettere dall'India (1912–1913)* (Milan, 1917).

Graham, Maria, *Journal of a Voyage to Brazil and Residence There During Part of the Years 1821, 1822, 1823* (London, 1824).

Hall, Charles Francis, *Arctic Researches and Life among the Esquimaux: Being the Narrative of an Expedition in Search of Sir John Franklin, in the Years 1860, 1861, and 1862* (New York, 1865).

Hand-book for Travellers in Northern Italy (London, 1847a, 3rd edition).

'Instructio Peregrinatoris quam sub praesidio D. D. Car. Linnaei proposuit Ericus And. Nordblad Gevalia-Gestricius Upsaliae 1759 Maji 9' in *Amoenitates Academicae: Volume V* (Erlangen, 1788, 2nd edition).

King, Samuel William *The Italian Valleys of the Pennine Alps* (London, 1858).

Kingsley, Mary, *Travels in West Africa, Congo Français, Corisco and Cameroons* (London, 1897).

Kingsley, Mary, *West African Studies* (London, 1901, 2nd edition).

Linnaeus, Carl, *Lachesis Lapponica or a Tour in Lapland: Volumes I and II* (London, 1811).

Morgan, Sydney, *Italy: voll. I-II* (Paris, 1821).

Nectoux, Hippolyte, *Observations sur la préparation des envois de plantes et arbres des Indes Orientales pour l'Amérique, et leur traitement pendant la traversée*, in *Mémoires d'agriculture, d'économie rurale et domestique publiés par la Société royale d'agriculture*, winter issue (1791), pp. 110–123.

North, Marianne, *Recollections of a Happy Life: Volumes I and II*, North, Janet Catherine (ed) (New York and London, 1892).

Nott, John, and Santi, Giorgio, *A Chemical Dissertation on the Thermal Waters of Pisa, and on the Neighbouring Acidolous Spring of Asciano with a Historical Sketch of Pisa, and a Meteorological Account of Its Weather to Which Are Added Analytical Papers Respecting the Sulphureous Waters of Yverdun* (London, 1793).

Palestine and Syria. Handbook for Travellers (Leipzig and London, 1876).

Piccolo, Francesco Maria, *Mémoire touchant l'éstat des missions nouvellement éstablies dans la Californie par les pères de la Compagnie de Jésus,* in *Lettres édifiantes et curieuses: Volume V* (Paris, 1705).

Pollini, Ciro, *Viaggio al Lago di Garda e al Monte Baldo in cui si ragiona delle cose naturali di quei luoghi aggiuntovi un cenno sulle curiosita del Bolca e degli altri monti veronesi* (Verona, 1816).

Radloff, Fredrik Wilhelm, *Beskrifning öfver Åland* (Åbo, 1795).

Raffles, Stamford (ed), *The Mission to Siam, and Hué the Capital of Cochin China,* in *the Years 1821–2, from the Journal of the Late George Finlayson, Esq.* (London, 1826).

Shelley, Percy Bysshe, and Wollstonecraft Shelley, Mary, *Notes to the Complete Poetical Works of Percy Bysshe Shelley* (London, 1839).

Smith, Elizabeth Bruce Elton, *The East India Sketch-Book: Comprising an Account of the Present State of Society in Calcutta, Bombay, etc.: Volume 2* (London, 1832).

Spallanzani, Lazzaro, *Viaggi alle Due Sicilie e in alcune parti dell'Appennino: Volume I* (Pavia, 1792).

The Life and Letters of Charles Darwin, Including an Autobiographical Chapter, in His Son, Francis Darwin (eds), *Three Volumes* (London, 1887).

The Life, Travels and Books of Alexander von Humboldt with an Introduction by Bayard Taylor (New York, 1859).

Volney, Constantin-François, *Tableau du climat et du sol des Etats-Unis d'Amérique: Volume I* (Paris, 1803).

Volney, Constantin-François, *Voyage en Syrie et en Egypte, pendant les années 1783, 1784 et 1785: Volumes I and II* (Paris, 1787, 2nd edition).

Volta, Giovanni Serafino, *Descrizione del Lago di Garda e de' suoi contorni con osservazioni di storia naturale e di belle arti* (Mantua, 1828).

von Humboldt, Alexander, *Essai sur la géographie des plantes accompagné d'un tableau physique der régions équinoxiales fondé sur des mesures exécutées […] pendant les années 1799, 1800, 1801, 1802 et 1803 par A. de Humboldt et A. Bonpland* (Paris, 1805).

von Humboldt, Alexander, *Researches Concerning the Institutions and Monuments of the Ancient Inhabitants of America with Descriptions and Views of Some of the Most Striking Scenes in the Cordilleras! Written in French by Alexander de Humboldt and Translated into English by Helen Maria Williams: Volume I* (London, 1814).

von Humboldt, Alexander, *Voyage aux régions équinoxiales du nouveau continent, fait en 1799, 1800, 1801, 1802, 1803 et 1804: Volume I* (Paris, 1816).

von Humboldt, Alexander, *Voyage aux régions équinoxiales du nouveau continent, fait en 1799, 1800, 1801, 1802, 1803 et 1804: Volume V* (Paris, 1820).

von Langsdorff, George Heinrich, *Voyages and Travels in Various Parts of the World, during the Years 1803, 1804, 1805, 1806, and 1807* (London, 1817).

Walker, Anna Maria, 'Journal of An Ascent to the Summit of Adam's Peak, Ceylon', *Companion to the Botanical Magazine,* Volume 1, pp 3–14. (1835).

Wilson, William (ed), *A Missionary Voyage to the Southern Pacific Ocean Performed in the Years 1796, 1797, 1798 in the Ship Duff…* (London, 1799).

Wollstonecraft Shelley, Mary, *Rambles in Germany and Italy in 1840, 1842, and 1843: Volumes I and II* (London, 1844).

Secondary sources

5 Nights: Wildlife, Glaciers & Cape Horn, https://www.swoop-patagonia.com/travel/cruises/wildlife-cape-horn, accessed on November 15, 2022.

Abbate Badin, Donatella, *Lady Morgan's Italy: Anglo-Irish Sensibility and Italian Realities* (Bethesda, 2007).

Acot, Pascal (ed), *The European Origins of Scientific Ecology (1800–1901)* (Amsterdam, 1998).

Adventure Cruise: Cape Horn & Glaciers, https://www.swoop-patagonia.com/adventure-cruise-cape-horn-glaciers, accessed on November 15, 2022.

Afolayan, Funso, *Culture and Customs of South Africa* (Westport CT and London, 2004).

Aiken, Edwin James, *Revd. Prof. Josias Leslie Porter 1823–1889*, in Lorimer, Hayden, Withers, Charles W.J. (eds), *Geographers: Biobibliographical Studies: Volume 26* (London and New York, 2007), pp. 67–78.

Akel, Regina, *Maria Graham: A Literary Biography* (Amherst and New York, 2009).

Akhimie, Patricia, and Andrea, Bernadette (eds), *Travel and Travail: Early Modern Women, English Drama, and the Wider World* (Lincoln and London, 2019).

al-Sayyid, Marsot, Afaf Lutfi, *Egypt in the Reign of Muhammad Ali* (Cambridge and New York, 1984).

Albrecht, Julia N. (ed), *Managing Visitor Experiences in Nature-Based Tourism* (Wallingford and Boston, 2021).

Alippi Cappelletti, Maurizia, 'Giglioli, Enrico Hillyer' *Dizionario Biografico degli Italiani: Volume 54*, https://www.treccani.it/enciclopedia/enrico-hillyer-giglioli_%28Dizionario-Biografico%29/ (Rome, 2000).

Amarasinghe, A.A. Thasun, Madawala, Majintha B., Karunarathna, D.M.S. Suranjan, Manolis, S. Charlie, de Silva, Anslem, and Sommerlad, Ralf, 'Human-crocodile Conflict and Conservation Implications of Saltwater Crocodiles *Crocodylus porosus* (Reptilia: Crocodylia: Crococodylidae) in Sri Lanka', *Journal of Threatened Taxa*, Volume 7, No. 5 (April 2015), pp. 7111–7130.

Ambrosoli, Mauro, *The Wild and the Sown: Botany and Agriculture in Western Europe: 1350–1850* (Cambridge and New York, 1997a).

Ambrosoli, Mauro, *The Wild and the Sown: Botany and Agriculture in Western Europe, 1350–1850* (Cambridge and New York, 1997b).

Anderson, Katharine, *Natural history and the scientific voyage*, in Curry, Helen Anne, Jardine, Nicholas, Secord, James Andrew, Spary, Emma C. (eds), *Worlds of Natural History* (Cambridge and New York, 2018a), pp. 304–318.

Anderson, Thomas J., *Reassembling the Strange: Naturalists, Missionaries, and the Environment of Nineteenth-Century Madagascar* (Lanham, Boulder, New York and London, 2018b).

Andersson, Gudrun, and Stobart, Jon, *Daily Lives and Daily Routines in the Long Eighteenth Century* (New York and Abingdon, 2022).

Andrades, Lidia, and Dimanche, Frederic, 'Destination Competitiveness and Tourism Development in Russia: Issues and Challenges', *Tourism Management*, Volume 62, pp. (October 2017). 360–376.

Anselmi, Francesco Antonio, *Sustainable Tourism Development: Ecotourism and Governance of Glocal Tourism* (Milan, 2020).

Apollo, Michal, and Andreychouk, Viacheslav, *Mountaineering, Adventure Tourism and Local Communities: Social, Environmental and Economics Interactions* (Cheltenham and Northampton MA, 2022).

Arato, Franco, 'Carlo Amoretti e il giornalismo scientifico nella Milano di fine Settecento', *Annali della Fondazione Luigi Einaudi*, Volume 21 (1987), pp. 175–216.

Archbold, William A.J., and Baigent, Elizabeth, *King, Samuel William*, in *Oxford Dictionary of National Biography*, https://doi.org/10.1093/ref:odnb/15598, accessed on December 2, 2022 (Oxford, 2004).

Archinard, Margarida, 'Les instruments scientifiques d'Horace-Bénédict de Saussure', *Le Monde alpin et rhodanien. Revue régionale d'ethnologie*, Volume 16, No. 1–2 (1988), pp. 151–164.

Arecco, Davide, *Mongolfiere, scienze e lumi nel tardo Settecento: Cultura accademica e conoscenze tecniche dalla vigilia della Rivoluzione francese all'età napoleonica* (Bari, 2003).

Argemí i d'Abadal, Lluis, and Lluch, Ernest (eds), *Agronomía y fisiocracia en España (1750–1820)* (Valencia, 1985).

Armiero, Marco, *Views from the South: Environmental Stories from the Mediterranean World (19th–20th Centuries)* (Naples, 2006).

Austin, Brian, *Marine Microbiology* (Cambridge and New York, 1988).

'Australia: Lost German tourist 'ate flies' to survive', *BBC News Services*, March 7, 2014, https://www.bbc.com/news/world-asia-26463979, accessed on October 21, 2022.

Baak, Lawrence J., *Undying Curiosity: Carsten Niebuhr and the Royal Danish Expedition to Arabia (1761–1767)* (Stuttgart, 2014).

Bainbridge, William, 'Titian Country: Josiah Gilbert (1814–1893) and the Dolomite Mountains', *Journal of Historical Geography*, Volume 56 (April 2017), pp. 22–42.

Balboni, Luigi Antonio, *Gl'italiani nella civiltà egiziana del secolo XIX. Storia – biografie – monografie: Volume I* (Alexandria, 1906).

Baldacchino, Godfrey (ed), *Extreme Tourism: Lessons from the World's Cold Water Islands* (Kidlington and Amsterdam, 2006).

Baldini, Ugo, *Borelli, Giovanni Alfonso*, in *Dizionario Biografico degli Italiani: Volume 12*, https://www.treccani.it/enciclopedia/giovanni-alfonso-borelli_%28 Dizionario-Biografico%29/, accessed on December 13, 2022 (Rome, 1971).

Baldini, Ugo, *Della Torre, Giovanni Maria*, in *Dizionario Biografico degli Italiani: Volume 37*, https://www.treccani.it/enciclopedia/della-torre-giovanni-maria_%28 Dizionario-Biografico%29/, accessed on December 13, 2022 (Rome, 1989).

Ballériaux, Catherine, *Missionary Strategies in the New World, 1610–1690: An Intellectual History* (New York and Abingdon, 2019).

Banfi, Enrico, Bianconi, Filiippo, Domenici, Davide, and Visconti, Agnese, *Making Alexander von Humboldt's Works Known in Lombardy: The Americna Volumes of Giulio Ferrari's* Costume antico e modern, monographic issue of *Natura*, Volume 111, No. 2 (2021).

Banfi, Enrico, and Visconti, Agnese, 'L'Orto di Brera alla fine della dominazione asburgica e durante l'età napoleonica', *Atti della Società Italiana di Scienze Naturali*, Volume II, No. 154 (2013), pp. 173–264.

Baraldi, Fulvio, 'Giovanni Serafino Volta, chimico, mineralogista e paleontologo mantovano (Mantova, 1754–1842)', *Atti e memorie – Accademia Nazionale Virgiliana di Scienze Lettere e Arti*, Volume 81 (2013), pp. 17–46.

Barman, Roderick J., 'The Forgotten Journey: Georg Heinrich Langsdorff and the Russian Imperial Scientific Expedition to Brazil, 1821–1829', *Terrae Incognitae*, Volume 3, No. 1 (1971), pp. 67–96.

Barry, Amanda, Cruickshank, Joanna, Brown-May, Andrew, and Grimshaw, Patricia (eds), *Evangelists of Empire? Missionaries in Colonial History* (Melbourne, 2008).

Bas, Martín Nicolás, and López Terrada, María Luz, *Una aproximación a la biblioteca del botánico valenciano Antonio José Cavanilles (1745–1804)*, in Real Sociedad Económica de Amigos del País (ed), *Antonio José Cavanilles (1745–1804): segundo centenario de la muerte de un gran botánico* (Valencia, 2004), pp. 201–285.

Battilani, Patrizia, *Vacanze di pochi, vacanze di tutti. L'evoluzione del turismo europeo* (Bologna, 2007).

Battistoni, Marco, *L'amministrazione sabauda e I traffici commerciali nel secolo XVIII*, in Torre, Angelo (ed), *Per vie di terra. Moveimenti di uomini e di cose nelle società di antico regime* (Milan, 2007), pp. 109–132.

Beattie, Andrew, *The Alps: A Cultural History* (Oxford and New York, 2006).

Beaurepaire, Pierre-Yves, *Les Lumières et le Monde : Voyager, Explorer, Collectionner* (Paris, 2019).

Becherelli, Alberto, *La politica adriatica e le Province Illiriche*, in Giovanna Motta (ed), *L'imperatore dei francesi e l'Europa napoleonica* (Rome, 2014), pp. 185–194.

Bennet, Betty T. (ed), *Selected Letters of Mary Wollstonecraft Shelley* (Baltimore, 1995).

Benson, Adolph B. (ed), *Peter Kalm's Travels in North America: The English Version of 1770: Volume 1* (New York, 1966).

Berdine, Michael D., *The Accidental Tourist, Wilfrid Scawen Blunt, and the British Invasion of Egypt in 1882* (Abingdon and New York, 2005).

Berger, Iris, *South Africa in World History* (Oxford and New York, 2009).

Berman, Jacob Rama, *American Arabesque: Arabs, Islam, and the Nineteenth-Century Imaginary* (New York and London, 2012).

Bernault, Florence, *Colonial Transactions: Imaginaries, Bodies, and Histories in Gabon* (Durham, NC, 2019).

Berti, Giampiero, *Un naturalista dall'Ancien Régime alla Restaurazione: Giambattista Brocchi (1772–1826)* (Bassano del Grappa, 1988).

Bhana, Surendra, and Brain, Joy B., *Setting Down Roots: Indian Migrants in South Africa, 1860–1911* (Johannesburg, 1990).

Bigatti, Giorgio (ed), *Quando l'Europa ci ammirava: viaggiatori, artisti, tecnici e agronomi stranieri nell'Italia del '700 e '800* (Truccazzano, 2016).

Bigger, Andreas, *Socin, Albert*, in *Historisches Lexikon der Schweiz (HLS)*, https://hls-dhs-dss.ch/de/articles/041667/2011-11-21/, accessed on December 16, 2022 (Basel, 2011).

Bindi, Umberto, *Giorgio Santi, scienziato pientino del Settecento. Biografia e scritti inediti* (Pienza, 2014).

Birkett, Dea, *Mary Kingsley: Imperial Adventuress* (London, 1992).

Black, Christopher F., *Early Modern Italy: a Social History* (London and New York, 2001).

Black, Jeremy, *The British and the Grand Tour* (Beckenham, 1985).

Black, Jeremy, *Natural and Necessary Enemies: Anglo-French Relations in the Eighteenth Century* (Athens GA, 1986).

Black, Jeremy, *France and the Grand Tour* (Basingstoke and New York, 2003).

Blair, Leon B., 'The Origin and Development of the Arabian Horse', *The Southwestern Historical Quarterly*, Volume 68, No. 3 (January 1965), pp. 303–316.

Bleichmar, Daniela, *Visible Empire: Botanical Expeditions and Visual Culture in the Hispanic Enlightenment* (Chicago and London, 2012).

Bleichmar, Daniela, *Visual Voyages: Images of Latin American Nature from Columbus to Darwin* (New Haven CT and London, 2017).

Bleichmar, Daniela, *Botanical Conquistadores*, in Curry, Helen Anne, Jardine, Nicholas, Secord, James Andrew, Spary, Emma C. (eds), *Worlds of Natural History* (Cambridge and New York, 2018), pp. 236–254.

Blum, Hester, *Charles Francis Hall's Arctic researches*, in Mentz, Steve, Rojas, Martha Elena (eds), *The Sea and Nineteenth-Century Anglophone Literary Culture* (Abingdon and New York, 2017), pp. 47–65.

Blunt, Alison, *Travel, Gender, and Imperialism: Mary Kingsley and West Africa* (New York, 1994).

Bollettino ufficiale del Ministero della Pubblica Istruzione, Volume 3, No. 7–8 (July–August 1877).

Borsarelli, Carmen, *La fortezza di Radicofani*, in Leonardo Rombai (ed), *I Medici e lo Stato senese: 1555–1609 storia e territorio* (Rome and Grosseto, 1980), pp. 133–143.

Boscani Leoni, Simona, Baumgartner, Sarah, and Knittel, Meike (eds), *Connecting Territories: Exploring People and Nature, 1700–1850* (Leiden and Boston, 2022).

Bourzes, Louis-Nöel, *Lettre au père Éstienne Souciet*, in *Lettres édifiantes et curieuses: Volume IX* (Paris, 1711), pp. 359–375.

Bray, Francesca, Hahn, Barbara, Lourdusamy, John Bosco, and Saraiva, Tiago, *Moving Crops and the Scales of History* (New Haven, 2023).

Brewer, John, 'Visiting Vesuvius: Guides, Local Knowledge, Sublime Tourism, and Science, 1760–1890', *The Journal of Modern History*, Volume 93, No. 1 (March 2021), pp. 1–33.

Briley-Webb, Linda, *Practical ocean navigation*, in Welch, Rosanne, Lamphier, Peg A. (eds), *Technical Innovation in American History: An Encyclopedia of Science and Technology: Volume 1* (Santa Barbara and Denver, 2019), pp. 137–138.

Brilli, Attilio, and Neri, Simonetta, *Le viaggiatrici del Grand Tour: storie, amori, avventure* (Bologna, 2020).

Brister, Lori, *The precise and the subjective: The guidebook industry and women's travel writing in late nineteenth-century Europe and North Africa*, in Clare Broome Saunders (ed), *Women, Travel Writing, and Truth* (New York and Abingdon, 2014), pp. 61–76.

Broc, Numa (ed), *Autour du monde: voyage de François de Pagès par terre et par mer, 1767–1771* (Paris, 1991).

Brockedon, William, *Illustrations of the Passes of the Alps, by which Italy Communicates with France, Switzerland, and Germany: Volume II* (London, 1829).

Brockmann, Sophie, *The Science of Useful Nature in Central America: Landscapes, Networks and Practical Enlightenment, 1784–1838* (Cambridge and New York, 2020).

Brockway, Lucile H. *Science and Colonial Expansion: The Role of the British Royal Botanic Gardens* (New York and London, 1979).

Bronchado, Ana, 'Nature-based Experiences in Tree Houses: Guests' Online Reviews', *Tourism Review*, Volume 74, No. 3 (2019), pp. 310–326.

Brooking, Tom, *The History of New Zealand* (Westport CT and London, 2004).

Brooks, Albert J., 'Elijah Hinsdale Burritt: The Forgotten Astronomer', *Popular Astronomy*, Volume 44 (1936), pp. 293–298.

Brotto-Pastega, Agostino, *Le case ed il Giardino del naturalista bassanese Alberto Parolini (1788–1867)* (Bassano del Grappa, 1988).

Browne, Janet, *Missionaries and the human mind: Charles Darwin and Robert Fitzroy*, in MacLeod, Roy, Rehbock, Philip F. (eds), *Darwin's Laboratory: Evolutionary Theory and Natural History in the Pacific* (Honolulu, 1994), pp. 263–282.

Browne, Janet, 'Making Darwin: Biography and the Changing Representations of Charles Darwin', *The Journal of Interdisciplinary History*, Volume 40, No. 3 (Winter 2010), pp. 347–373.

Bruce, David M., *The nineteenth-century "golden age" of cultural tourism: how the beaten track of the intellectuals became the modern tourism trail*, in Smith, Melanie, Richards, Greg (eds), *The Routledge Handbook of Cultural Tourism* (Abingdon and New York, 2013), pp. 11–18.

Bruce, Julia, 'Banks and Breadfruit', *RSA Journal*, Volume 141, No. 5444 (1993), pp. 817–820.

Buccieri, Giuseppina, *Gioeni, Giuseppe*, in *Dizionario Biografico degli Italiani: Volume 55* (Rome, 2001).

Busnardo, Giuseppe, *Alberto Parolini e la storia naturale del suo tempo*, in Alessandro Minelli (ed), *Storia naturale a Bassano (1788–1988)* (Padua, 1990), pp. 13–31.

Bussby, Frederick, 'John Nott, M.D. 1751–1825: The Oriental Travels of an Eighteenth-century Physician', *Medical History*, Volume 18, No. 3 (July 1974), pp. 294–298.

Busset, Thomas, Lorenzetti, Luigi, and Mathieu, Jon (eds), *Tourisme et Changements Culturels – Tourismus und Kulturelle Wandel* (Zürich, 2004).

Byrnes, Giselle (ed), *The New Oxford History of New Zealand* (Oxford and New York, 2009).

Caballero, M. Soledad, *Clashing tastes: European femininity and race in Maria Graham's journal of a voyage to Brazil*, in Cabañas, Miguel A., Dubino, Jeanne, Salles-Reese, Veronica, Totten, Gary (eds), *Politics, Identity, and Mobility in Travel Writing* (Abingdon and New York, 2015).

Campón-Cerro, Ana María, Hernández-Mogollón, José Manuel, Baptista Alves, Helena María, and Di-Clemente, Elide, *The tourist in rural destinations: an experiential approach based on relationships with local people and surroundings*, in Kastenholz, Elisabeth, João Carneiro, Maria, Eusébio, Celeste, Figueiredo, Elisabete (eds), *Meeting Challenges for Rural Tourism through Co-Creation of Sustainable Tourist Experiences* (Newcastle upon Tyne, 2016), pp. 103–131.

Caradonna, Jeremy L. (ed), *Routledge Handbook of the History of Sustainability* (Abingdon and New York, 2018).

Carandini, Leonardo, 'La Posta di Radicofani', *L'Universo*, Volume 44, No. 1 (1964), pp. 153–176.

Carassale, Alessandro, *"Cedri e palme all'hebrea". Produzione e commercio nell'estremo Ponente ligure tra XVI e XVIII secolo*, in Carassale, Alessandro, Littardi, Claudio (eds), *Frontiera Judaica. Gli ebrei nello spazio ligure-provenzale dal medioevo alla Shoah* (Saluzzo, 2021), pp. 101–127.

Carassale, Alessandro, Gandolfi, Daniela, and Guglielmi Manzoni, Alberto (eds), *Il viaggio in Riviera. Presenze straniere nel Ponente ligure dal XVI al XX secolo* (Bordighera, 2015).

Carozzi, Albert V., *Horace-Bénédict de Saussure (1740–1799) : un pionnier des sciences de la terre* (Geneva, 2005).

Cashdollar, Charles D., *The Transformation of Theology, 1830–1890: Positivism and Protestant Thought in Britain and America* (Princeton, 1989).

Cavallo, Sandra, and Storey, Tessa (eds), *Conserving Health in Early Modern Culture: Bodies and Environments in Italy and England* (Manchester, 2017).

Cawely, Mary, and Gillmor, Desmond A., 'Integrated Rural Tourism: Concepts and Practice', *Annals of Tourism Research*, Volume 35, No. 2 (2008), pp. 316–337.

Cencini, Carlo, *Il paesaggio naturale: i valori culturali della natura*, in Zerbi, Maria Chiara (ed), *Il paesaggio rurale: un approccio patrimoniale* (Turin, 2007), pp. 27–45.

Cerveny, Lee K., *Tourism and Its Effects on Southeast Alaska Communities and Resources: Case Studies from Haines, Craig, and Hoonah, Alaska*, (Washington DC, 2005).

Cerveny, Lee K., *Sociocultural Effects of Tourism in Hoonah Alaska*, (Washington DC, 2007).

Chaney, Edward, *The Evolution of the Grand Tour* (London, 1998).

Chen, Joseph S., and Prebensen, Nina K. (eds), *Nature Tourism* (Abingdon and New York, 2017).

Christou, Prokopis A., *The History and Evolution of Tourism* (Wallingford and Boston, 2022).

Ciancio, Luca, *Autopsie della Terra. Illuminismo e geologia in Alberto Fortis (1741–1803)* (Florence, 1995).

Ciardi, Marco (ed), *Esplorazioni e viaggi scientifici nel Settecento* (Milan, 2008).

Cimino, Guido, *De Filippi, Filippo*, in *Dizionario Biografico degli Italiani: Volume 33*, https://www.treccani.it/enciclopedia/filippo-de-filippi_(Dizionario-Biografico)/ (Rome, 1987).

Clark, Steve, and Smethurst, Paul (eds), *Asian Crossings: Travel Writing on China, Japan and Southeast Asia* (Aberdeen HK, 2008).

Cochrane, Janet, *Responses to Continuing Crisis in Sri Lanka*, in Cochrane, Janet (ed), *Asian Tourism: Growth and Change* (Oxford and Amsterdam, 2008), pp. 79–91.

Cohen, Ed, *A Body Worth Defending: Immunity, Biopolitics, and the Apotheosis of the Modern Body* (Durham and London, 2009).

Collier, Simon, and Sater, William F., *A History of Chile, 1808–2002* (Cambridge and New York, 2004, 2nd edition).

Colombo, Emanuele, *Piccolo, Francesco Maria*, in *Dizionario Biografico degli Italiani: Volume 83* (Rome, 2015).

Conca, Messina, Silvia, A. (ed), *Leading the Economic Risorgimento: Lombardy in the 19th Century* (Abingdon and New York, 2022).

Cook, Harold J., *Risking private ventures: the instructive failure of a well-traveled artist, Cornelis de Bruyn*, in Dijksterhuis, Fokko Jan (ed), *Regulating Knowledge in an Entangled World* (Abingdon and New York, 2022), pp. 169–192.

Cooper, Alix, *Inventing the Indigenous: Local Knowledge and Natural History in Early Modern Europe* (Cambridge and New York, 2007).

Cottam, Stella, and Orchiston, Wayne, *Eclipses, Transits, and Comets of the Nineteenth Century: How America's Perception of the Skies Changed* (Cham, 2015).

Cramsie, Elizabeth, 'Two Victorian Tourists Found Alive after Getting Lost Bushwalking in North Queensland', *ABC News*, May 22, 2021, https://www.abc.net.au/news/2021-05-22/bushwalkers-victoria-missing-north-queensland-search-mareeba/100157784, accessed on October 21, 2022.

Crick, Malcom, *Resplendent Sites, Discordant Voices: Sri Lankas and International Tourism* (Abingdon and New York, 1994).

'Crocodile attacks kills police officer in Matara', *News 1st*, July 2, 2020, https://www.newsfirst.lk/2020/07/02/crocodile-attacks-kills-police-officer-in-matara/, accessed on November 7, 2021.

Crosby, Alfred W., *The Columbian Exchange: Biological and Cultural Consequences of 1492* (Westport CT, 1972).

Crosby, Alfred W., *Ecological Imperialism: The Biological Expansion of Europe, 900–1900* (Cambridge and New York, 2004, 2nd edition).

Crouzet, François, *Britain Ascendant: Comparative Studies in Franco-British Economic History* (Cambridge and New York, 1990).

Cutter, Donald D., *Introduction*, in David, Andrew, Fernandez-Armesto, Felipe, Novi, Carlos, Williams, Glyndwr (eds), *The Malaspina Expedition 178–1794: The Journal of the Voyage by Alessandro Malaspina: Volume 1* (London and Madrid, 2001), pp. xxix–lxxvii.

Daniel, Carey, *Advice on the art of travel*, in Das, Nandini, Youngs, Tim (eds), *The Cambridge History of Travel Writing* (Cambridge and New York, 2019), pp. 392–407.

Das, Nandini, and Youngs, Tim (eds), *The Cambridge History of Travel Writing* (Cambridge and New York, 2019).

Daum, Andreas W., *German Naturalists in the Pacific around 1800: Entanglement, Autonomy, and a Transnational Culture of Expertise*, in Berghoff, Hartmut, Biess, Frank, Strasser, Ulrike (eds), *Explorations and Entanglements: Germans in Pacific Worlds from the Early Modern Period to World War I* (New York, 2019), pp. 70–102.

Davidann, Jon Thares, *The Limits of Westernization: American and East Asian Intellectuals Create Modernity, 1860–1960* (Abingdon and New York, 2019).

De Caprio, Vincenzo, 'Sul paesaggio finlandese e l'idea di natura in Giuseppe Acerbi', *Atti e memorie – Accademia Nazionale Virgiliana di Scienze Lettere e Arti*, Volume 85 (2017), pp. 185–206.

De Felice, Renzo, *Amoretti, Carlo*, in *Dizionario Biografico degli Italiani: Volume 3*, https://www.treccani.it/enciclopedia/carlo-amoretti_%28Dizionario-Biografico%29/, accessed on December 2022 (Rome, 1961).

De Frenza, Lucia, *I sonnambuli delle miniere: Amoretti, Fortis, Spallanzani e il dibattito sull'elettrometria organica e minerale in Italia (1790–1816)* (Florence, 2005).

de Hond, Jan, *Cornelis de Bruijn (1652–1726/27): a Dutch painter in the East*, in van Gelder, Geert Jan, de Moor, Ed(eds), *Eastward Bound: Dutch Ventures and Adventures in the Middle East* (Amsterdam and Atlanta, 1994), pp. 51–80.

de Silva, Anslem, 'Crocodiles: Our Living Dinosaurs (With Notes on the *kimbul kotuwa* or the Crocodile Excluding Enclosures)', *Loris: Journal of the Wildlife and Nature Protection Society of Sri Lanka*, Volume 27, No. 5–6 (December 2016), pp. 22–27.

de Souza, Luis Henrique, Kastenholz, Elisabeth, and de Lourdes de Azevedo Barbosa, Maria, 'Relevant Dimensions of Tourist Experiences in Unique, Alternative Person-to-person Accommodation—Sharing Castles, Treehouses, Windmills, Houseboats or House-buses', *International Journal of Hospitality & Tourism Administration*, Volume 21, No. 4 (2020), pp. 390–421.

del Arco Aguilar, Marcelino J., and Rodríguez Delgado, Octavio, *Vegetation of the Canary Islands* (Cham, 2018).

Del Soldato, Marco, 'Un'indagine giudiziale rivelatrice di tecniche minerarie secentesche in Valle Anzasca', *Bollettino Storico per la Provincia di Novara*, Volume 77, No. 2 (1986), pp. 111–126.

Dell'assoluta influenza delle foreste sulla temperatura atmosferica, in di Bérenger, Adolfo (ed), *Giornale di economia forestale ossia raccolta di memorie lette nel R. Istituto Forestale di Vallombrosa: Volume I* (Florence, 1871–1872), p. 60.

Desmond, Rey, *Kew: The History of the Royal Botanic Gardens* (London, 2007, 2nd edition).

Díaz, Dorismel, 'Charles Darwin and the Representation of Black Communities in His Travel Narrative', *Hallazgos*, Volume 12, No. 23 (2015), pp. 231–249.

Dilk, Enrica Yvonne, *Un agronomo austriaco nel Lombardo-Veneto*, in Burger, Johann, *Agricoltura del Regno Lombardo-Veneto: versione italiana del dottor V. P. con note del Dottor Giuseppe Moretti* [anastatic reprint of the 1843 Italian edition] (Milan, 2002), pp. 11–32.

Dolomites UNESCO World Heritage, https://www.dolomitiunesco.info/, accessed on December 2, 2022.

Donovan, Julie, *Sydney Owenson, Lady Morgan and the Politics of Style* (Bethesda, Dublin and Palo Alto, 2009).

Douglas, Gina, *Italy in the Linnaean collections: a review of the biological specimens, books and manuscripts linking Italy with Linnaeus and the Linnaean society of London*, in Beretta, Marco, Tosi, Alessandro (eds), *Linnaeus in Italy: The Spread of a Revolution in Science* (Sagamore Beach, 2007), pp. 31–46.

Dovers, Stephen, Edgecombe, Ruth, and Guest, Bill (eds), *South Africa's Environmental History: Cases and Comparisons* (Athens OH, 2003).

Dröscher, Ariane, *Plants and Politics in Padua During the Age of Revolution, 1820–1848* (Cham, 2021).

East, Georgia, *Wines of the Cape West Coast*, January 27, 2020, https://www.timeslive.co.za/sunday-times/lifestyle/food/2020-01-27-wines-of-the-cape-west-coast/, accessed on November 15, 2022.

Echeverri, Alejandra, Smith, Jeffrey R., MacArthur-Waltz, Dylan, and Daily, Gretchen C., 'Biodiversity and Infrastructure Interact to Drive Tourism to and Within Costa Rica', *The Proceedings of the National Academy of Sciences*, Volume 119, No. 11 (March 2022), https://doi.org/10.1073/pnas.2107662119.

Edmundson, William, *A History of the British Presence in Chile: From Bloody Mary to Charles Darwin and the Decline of British Influence* (New York, 2009).

Eliasson, Pår, *Humboldt, Alexander von*, in Jennifer Speake (ed), *Literature of Travel and Exploration: An Encyclopedia: Volume I* (Abingdon and New York, 2003), pp. 572–574.

Etheridge, Kay, *The history and influence of Maria Sibylla Merian's bird eating tarantula: Circulating images and the production of natural knowledge*, in Manning, Patrick, and Rood, Daniel (eds), *Global Scientific Practice* (Pittsburgh, 2016), pp. 54–70.

Etheridge, Kay, *The Flowering of Ecology: Maria Sibylla Merian's Caterpillar Book* (Leiden and Boston, 2021).

Evans, Joan, *The Conways: A History of Three Generations* (London, 1966).

Fagnani, Martino Lorenzo, 'From "Pure Botany" to "Economic Botany" – Changing Ideas by Exchanging Plants: Spain and Italy in the Late Eighteenth and the Early Nineteenth Century', *History of European Ideas*, Volume 48, No. 4 (2022a), pp. 402–420.

Fagnani, Martino Lorenzo, 'Travels and Representations at the Core of Western Agricultural Science: Discovering Rural Societies in Spain, Italy and Lebanon, Late Eighteenth and Early Nineteenth Centuries', *Continuity and Change*, Volume 37, No. 3 (December 2022b), p. 37.

Fagnani, Martino Lorenzo, *The Development of Agricultural Science in Northern Italy in the Late Eighteenth and Early Nineteenth Century* (Cham, 2023).

Fahmy, Khaled, *All the Pasha's Men: Mehmed Ali, His Army and the Making of Modern Egypt* (Cambridge and New York, 1997).

Fahmy, Khaled, *Mehmed Ali: From Ottoman Governor to Ruler of Egypt* (London, 2009).

Falk, Gregor C., Strecker, Manfred R., and Schneider Simon (eds), *Alexander von Humboldt: Multiperspective Approaches* (Cham, 2022).

Fazi, André, 'Volney et la Corse', *Bulletin de la Société des Sciences Historiques et Naturelles de la Corse*, No. 718–719 (2007), pp. 27–95.

Federico, Giovanni, *An Economic History of the Silk Industry, 1830–1930* (Cambridge and New York, 1997).

Fedi, Pierfrancesco, 'Alcune note su Giovanni De Riseis', *Rivista degli studi orientali*, Volume 78, No. 3–4 (2005), pp. 237–259.

Feldaman, Paula R., and Scott-Kilvert, Diana (eds), *The Journals of Mary Shelley 1814–1844: Volume 1* (Oxford, 1987).

Fennell, David A., *Ecotourism* (Abingdon and New York, 2020, 5th edition).

Finnegan, Rachel, *Richard Pococke's Letters from the East (1737–1740)* (Leiden, 2020).

Fleming, Fergus, *Killing Dragons: The Conquest of the Alps* (New York, 2000).

Foliard, Daniel, *Dislocating the Orient: British Maps and the Making of the Middle East, 1854–1821* (Chicago and London, 2017).

Food and Agriculture Organization of the United Nations (ed), *Sorghum and Millets in Human Nutrition* (Rome, 1995).

Frank, Katherine, *A Voyager Out: The Life of Mary Kingsley* (New York, 1986).

Fredman, Peter, and Haukeland, Jan V. (eds), *Nordic Perspectives on Nature-Based Tourism: From Place-Based Resources to Value-Added Experiences* (Cheltenham UK and Northampton MA, 2021).

Fredman, Peter, and Tyrväinen, Liisa (eds), *Frontiers in Nature-based Tourism: Lessons from Finland, Iceland, Norway and Sweden* (Abingdon and New York, 2011).

Fressoz, Jean-Baptiste, and Locher, Fabien, *Les révoltes du ciel: une histoire du changement climatique XVᵉ-XXᵉ siècle* (Paris, 2020).

Frost, Robert I., *The Northern Wars. War, State and Society in Northeastern Europe, 1558–1721* (Harlow, 2000).

Fullerton, Ronald A., *The Foundations of Marketing Practice: A History of Book Marketing in Germany* (Abingdon and New York, 2016).

García Belmar, Antonio, and Bertomeu Sánchez, José Ramón, *Constructing the Centre from the Periphery*, in Simões, Ana, Carneiro, Ana, Diogo, Maria Paula (eds), *Travels of Learning: A Geography of Science in Europe* (Dordrecht, Boston and London, 2003), pp. 143–188.

Garnett, Richard, and Baigent, Elizabeth, *Browne, William George*, in *Oxford Dictionary of National Biography*, https://doi.org/10.1093/ref:odnb/3710, accessed on December 5, 2022 (Oxford, 2004).

Garrod, Brian, Wornell, Roz, and Youell, Ray, 'Re-conceptualising Rural Resources as Countryside Capital: The Case of Rural Tourism', *Journal of Rural Studies*, Volume 22, No. 1 (2006), pp. 117–128.

Gaulmier Jean, 'Note sur l'itinéraire de Volney en Égypte et en Syrie', *Bulletin d'Études Orientales*, Volume 13 (1949–1951), pp. 45–50.

Gee, Brian, *Francis Watkins and the Dollond Telescope Patent Controversy*, in Anita McConnell, A.D. Morrison-Low (eds) (Farnham and Burlington VT, 2014).

Georgi, Claudia, *Maria Graham, travel writing on India, Italy, Brazil, and Chile (1812–1824)*, in Schaff, Barbara (ed), *Handbook of British Travel Writing* (Berlin and Boston, 2020), pp. 313–334.

Gerassi-Navarro, Nina, *Women, Travel, and Science in Nineteenth-Century Americas: The Politics of Observations* (Cham, 2017).

Giacomini, Valerio, *Brocchi, Giovanni Battista*, in *Dizionario Biografico degli Italiani: Volume 14*, https://www.treccani.it/enciclopedia/giovanni-battista-brocchi_%28 Dizionario-Biografico%29/ (Roma, 1972).

Giannetto Pini, Annamaria, *In feluca sul Nilo: una crociera alternativa*, https://www.viaggionelmondo.net/25350-in-feluca-sul-nilo-crociera-alternativa/m accessed on November 14, 2022.

Gibbs, Jenna M. (ed), *Global Protestant Missions: Politics, Reform, and Communication, 1730s–1930s* (Abingdon and New York, 2020).

Gilbert, Josiah, and Churchill, George Cheetham, *The Dolomite Mountains. Excursions through Tyrol, Carinthia, Carniola, and Friuli in 1861, 1862, and 1863* (London, 1864).

Gilje, Paul A., *To Swear Like a Sailor. Maritime Culture in America, 1750–1850* (Cambridge and New York, 2016).

Gillespie, William M., *A Treatise on Surveying Comprising the Theory and the Practice… Revised and Enlarged by Cady Staley, Ph.D., President of Case School of Applied Science: Volume I* (New York, 1896).

Gleadhill, Emma, *Taking Travel Home: The Souvenir Culture of British Women Tourists, 1750–1830* (Manchester, 2022).

Glover, Michael, *A Very Slippery Fellow – The Life of Sir Robert Wilson 1777–1849* (Oxford, 1978).

Goldsmith, Sarah, *Masculinity and Danger on the Eighteenth-Century Grand Tour* (London, 2020).

Gondoli, Antonio, and Natali, Antonio (eds), *Luoghi della Toscana medicea* (Florence, 1980).

González Bueno, Antonio, *Gómez Ortega, Cavanilles, Zea, tres botánicos de la Ilustración: la ciencia al servicio del poder* (Madrid, 2002).

González Bueno, Antonio, and Rodríguez Nozal, Raúl, *Plantas americanas para la España ilustrada: génesis, desarrollo y ocaso del proyecto español de expediciones botánicas* (Madrid, 2000).

Goodwin, Gráinne, and Johnston, Gordon, 'Guidebook Publishing in the Nineteenth Century: John Murray's *Handbooks for Travellers*', *Studies in Travel Writing*, Volume 17, No. 1 (2013), pp. 43–61.

Gray, Christopher J., *Colonial Rule and Crisis in Equatorial Africa: Southern Gabon, c. 1850–1940* (Rochester NY, 2002).

Grosvenor Myer, Valerie, *A Victorian Lady in Africa: The Story of Mary Kingsley* (London, 1989).

Grove, Jean M., *Little Ice Ages: Ancient and Modern: Volume I* (London and New York, 2004, 2nd edition).

Grusin, Richard, *Culture, Technology, and the Creation of America's National Parks* (Cambridge and New York, 2004).

Guglielminetti, Marziano, *Gozzano, Guido*, in *Dizionario Biografico degli Italiani: Volume 58*, https://www.treccani.it/enciclopedia/guido-gozzano_%28Dizionario-Biografico%29/, accessed on December 6, 2022. (Rome, 2002).

Gunawardhana, A.W., 'Crocodile Kills Man in Nilwala Ganga', *Daily News. Sri Lanka's National Newspaper*, September 5, 2007, http://archives.dailynews.lk/2007/09/05/news16.asp, accessed on October 21, 2022.

Gunn, Mary, and Codd, Leslie E., *Botanical Exploration of Southern Africa* (Cape Town and Pretoria, 1981).

Hackl, Franz, Halla, Martin, and Pruckner, Gerald J., 'Local Compensation Payments for Agri-environmental Externalities: A Panel Data Analysis of Bargaining Outcomes', *European Review of Agricultural Economics*, Volume 34, No. 3 (2007), pp. 295–320.

Halevi, Sharon, 'In Sunshine and in Shadow: Adolescent Girls and Thanatourism in the Early American Republic', *Journal of Tourism History*, Volume 12, No. 1 (2020), pp. 71–85.

Hall, Derek R., Roberts, Lesley, and Morag, Mitchell (eds), *New Directions in Rural Tourism* (Hants, 2005, 2nd edition).

Hall, Michael C., and Boyd, Stephen (eds), *Nature-Based Tourism in Peripheral Areas: Development or Disaster?* (Clevedon, Buffalo and Toronto, 2005).

Hand-book for Travellers in Northern Italy (London, 1847b, 3rd edition).

Hansen, Peter H., *Conway, (William) Martin, Baron Conway of Allington'*, in *Oxford Dictionary of National Biography*, https://doi.org/10.1093/ref:odnb/32536, accessed on December 2, 2022 (Oxford, 2004).

Hansen, Peter H., *The Summits of Modern Man: Mountaineering after the Enlightenment* (Cambridge MA, 2013).

Hanson, Paul R., *Historical Dictionary of the French Revolution* (Lanham, Toronto and Oxford, 2004).

Haouas, Dalila, Mdellel, Lassaad, Mraihi, I., Hafsi, C., and Balmès, V., '*Pollinia pollini* (Costa, 1857) (Hemiptera, Asterolecaniidae) Infesting Olive Trees: A First Record in Tunisia', *Bulletin OEPP – EPPO Bulletin*, Volume 50, No. 1 (2020), pp. 201–202.

Hart, Darryl G., *Reformed theology and global Christianity: The cases of South Africa and Korea*, in Allen, Michael, Swain, Scott R. (eds), *The Oxford Handbook of Reformed Theology* (Oxford and New York), pp. 171–186.

Harvey, Robert, *The War of Wars: The Great European Conflict 1793–1815* (New York, 2007).

Hayden, Judy A. (ed), *Travel Narratives, The New Science, and Literary Discourse, 1569–1750* (Farnham and Burlington VT, 2012).

Higgitt, Rebekah, *Equipping expeditionary astronomers: Nevil Maskelyne and the development of "Precision Exploration"*, in MacDonald, Fraser, Withers, Charles W.J. (eds), *Geography, Technology and Instruments of Exploration* (London and New York, 2015), pp. 15–36.

Holden, Andrew, and Fennell, David (eds), *The Routledge Handbook of Tourism and Environment* (Abingdon and New York, 2013).

Holland, Clive, *Franklin, Sir John*, in Halpenny, Francess G. (ed), *Dictionary of Canadian Biography: Volume 7* (Toronto, 1988), pp. 323–328.

Hughes, J. Donald, *What is Environmental History?* (Malden MA, 2006).

Hunt, Carter A., Durham, William H., Driscoll, Laura, and Honey, Martha, 'Can Ecotourism Deliver Real Economic, Social, and Environmental Benefits? A study of the Osa Peninsula, Costa Rica', *Journal of Sustainable Tourism*, Volume 23, No. 3 (2015), pp. 339–357.

Hunt, Holly, and Rogerson, Christian M., *Tourism-led development and backward linkages: evidence from the agriculture-tourism nexus in Southern Africa*, in Visser, Gustav, Ferreira, Sanette (eds), *Tourist and Crisis* (Abingdon and New York, 2013), pp. 159–179.

Il Movimento Cattolico, October 15, 1885.

Ireland, Aideen, 'Richard Pococke (1704–65), Antiquarian', *Peritia: Journal of Medieval Academy of Ireland*, Volume 20 (2008), pp. 353–378.

Isenberg, Andrew C. (ed), *The Oxford Handbook of Environmental History* (Oxford, 2014).

Italian Senate: Giovanni De Riseis: https://notes9.senato.it/web/senregno.nsf/c1544f301fd4af96c125785d00598476/7439e0ab914cdaab4125646f005b0c8e?OpenDocument, accessed on October 19, 2022.

Ito, Takatoshi, and Hoshi, Takeo, *The Japanese Economy* (Cambridge MA and London, 2020, 2nd edition).

James, Gregory, *Tamil Lexicography* (Tübingen, 1991).

Jansen, Ena, *Like Family: Domestic Workers in South African History and Literature* (Johannesburg, 2019).

Jefferson, Thomas, *Memorandums taken on a journey from Paris into the southern parts of France and Northern Italy, in the year 1787*, in Thomas Jefferson Randolph (ed), *Memoirs, Correspondence, and Private Papers of Thomas Jefferson, Late President of the United States: Volume II* (London, 1829), pp. 115–160.

Jensen, Øystein, Lindberg, Frank, Kieti, Damiannah, M. Bjørn, Åmo, Willy, and Nampushi, James S., *How local traditions and way of living influence tourism: basecamp explorer in Maasai Mara, Kenya and Svalbard, Norway*, in Chen, Joseph S., Prebsen, Nina K. (eds), *Nature Tourism* (Abingdon and New York, 2017), pp. 68–81.

Johnston, Judy, *Victorian Women and the Economies of Travel, Translation, and Culture, 1830–1870* (Abingdon and New York, 2016, 2nd edition).

Jones, Peter M., *Agricultural Enlightenment: Knowledge, Technology, and Nature, 1750–1840* (Oxford and New York, 2016).

Jurriaans-Helle, Geralda, *Cornelis De Bruijn: voyages from Rome to Jerusalem and from Moscow to Batavia* (Amsterdam, 1998).

Keighren, Innes M., Withers, Charles W. J., and Bell, Bill, *Travels into Print: Exploration, Writing, and Publishing with John Murray, 1773–1859* (Chicago and London, 2013).

Keynes, Richard Darwin (ed), *The Beagle Record: Selections from the Original Pictorial Records and Written Accounts of the Voyage of H.M.S. Beagle* (Cambridge and New York, 1979).

Kingsland, Sharon E., *The Evolution of American Ecology, 1890–2000* (Baltimore, 2005).

Kiple, Kenneth F., *A Movable Feast: Ten Millennia of Food Globalization* (Cambridge and New York, 2007).

Kloppenberg, James T., *Toward Democracy: The Struggle for Self-Rule in European and American Thought* (Oxford and New York, 2016).

Koerner, Lisbet, *Carl Linnaeus in his time and place*, in Jardine, Nicholas, Secord, James Andrew, Spary, Emma C. (eds), *Cultures of Natural History* (Cambridge, 1996), pp. 145–177.

Kohli, Atul, *Imperialism and the Developing World: How Britain and the United States Shaped the Global Periphery* (Oxford and New York, 2020).

Kuenzi, Caroline, and McNeely, Jeff, *Nature-based tourism'*, in Renn, Ortwin, Walker, Katherine D. (eds), *Global Risk Governance: Concept and Practice Using the IRGC Framework* (Dordrecht, 2008), pp. 155–178.

L'Avvisatore Mercantile: Giornale di commercio e industria, January 17 (1852), p. 14.

La Mela, Matti, 'Property Rights in Conflict: Wild Berry-picking and the Nordic Tradition of *allemansrätt*', *Scandinavian Economic History Review*, Volume 62, No. 3 (2014), pp. 266–289.

Laband, John, *The Transvaal Rebellion: The First Boer War, 1880–1881* (Abingdon and New York, 2014).

Laity, Julie J., *Deserts and Desert Environments* (New York, 2009).

Lambert, Andrew, *Franklin: Tragic Hero of Polar Navigation* (London, 2009).

Land, Isaac, *War, Nationalism, and the British Sailor, 1750–1850* (New York, 2009).

Lappo, Georgiĭ, *Arkhangel'sk*, in Nuttall, Mark (ed), *Encyclopedia of the Arctic: Volumes 1, 2 and 3* (New York and Abingdon, 2005), pp. 149–150.

Laureti, Lamberto, *Italian contributions during the time of Werner relating to plutonism and neptunism – The works of Esprit-Benoit Nicolis de Robilant and Scipione Breislak*, in Albrecht, Helmuth, Ladwig, Roland (eds), *Abraham Gottlob Werner and the Foundation of the Geological Sciences. Selected Papers of the International Werner Symposium in Freiberg, 19th–24th September 1999* (Freiberg, 2003), pp. 179–187.

Lazarus, Maureen H., and Pardoe, Heather S. (eds), *Catalogue of Botanical Prints and Drawings at the National Museums & Galleries of Wales* (Cardiff, 2003).

Le-May Sheffield, Suzanne, *Revealing New Worlds: Three Victorian Women Naturalists* (London and New York, 2001).

Leask, Nigel, *Curiosity and the Aesthetics of Travel-Writing, 1770–1840* (Oxford and New York, 2002).

Leddra, Michael, *Time Matters: Geology's Legacy to Scientific Thought* (Oxford, Chichester and Hoboken, 2010).

Leland Crosthwait, Herbert, 'A Journey to Lake Saint Martin, Patagonia', *The Geographical Journal*, Volume 25, No. 3 (1905), pp. 286–291.

Lennon, John, and Foley, Malcolm, *Dark Tourism: the Attraction of Death and Disaster* (London, 2000).

Levati, Stefano, and Liva, Giovanni, *Viaggio di quasi tutta l'Europa colle viste del commercio dell'istruzione e della salute. Lettere di Paolo e Giacomo Greppi al padre (1777–1781)* (Milan-Cinisello Balsamo, 2006), pp. 14–58.

Libo, Yan, 'Origins of Nature Tourism in Imperial China', *Journal of Tourism Futures*, Volume 4, No. 3 (2018), pp. 265–274.

Lindskog, Annika, *Constructing and classifying "The North": Linnaeus and Lapland*, in Duffy, Cian (ed), *Romantic Norths: Anglo-Nordic Exchanges, 1770–1842* (Cham, 2017), pp. 75–99.

Londt, Leigh-Ann, *From grape vines to yummy food, Groote Post never disappoints*, May 11, 2021, https://www.capetownetc.com/cape-town/restaurants/from-grape-vines-to-yummy-food-groote-post-never-disappoints-ready/, accessed on November 15, 2022.

Longhurst, Alan, *Ecological Geography of the Sea* (Burlington, San Diego and London, 2007, 2nd edition).

Loomis, Chauncey C., *Weird and Tragic Shores: The Story of Charles Francis Hall* (New York, 2000).

Louise, Mary, *Imperial Eyes: Travel Writing and Transculturation* (London and New York, 1991).

Lumbroso, Giacomo, *Descrittori italiani dell'Egitto e di Alessandria* (Rome, 1879).

Lunario genovese compilato dal signor Regina & C. – Guida amministrativa e commerciale di Genova e provincia (Genoa, 1887).

MacDonald, Fraser, and Withers, Charles W.J. (eds), *Geography, Technology and Instruments of Exploration* (London and New York, 2015).

Madriñan, Santiago, *Nikolaus Joseph Jacquin's American Plants Botanical Expedition to the Caribbean (1754–1759) and the Publication of the Selectarum Stirpium Americanarum Historia* (Leiden and Boston, 2013).

Maffi, Luciano, *The traveling priest: food for the spirit and food for the body*, in Williot, Jean Pierre, Blanquis, Isabelle (eds), *Nomadic Food: Anthropological and Historical Studies around the World* (Lanham and London, 2019), pp. 147–168.

Maffi, Luciano, and Fagnani, Martino Lorenzo, *Disability and Tourism in Nineteenth- and Twentieth-Century Italy* (Abingdon and New York, 2021).

Magra, Christopher, *Chocolate and the Atlantic economy: circuits of trade and knowledge*', in Victoria Barnett-Woods (ed), *Cultural Economies of the Atlantic World: Objects and Capital in the Transatlantic Imagination* (New York and Abingdon, 2020), pp. 173–190.

Manfredi, Dario, *Malaspina, Alessandro*, in *Dizionario Biografico degli Italiani: Volume 67*, https://www.treccani.it/enciclopedia/alessandro-malaspina_%28 Dizionario-Biografico%29/ (Rome, 2006).

Margaryan, Lusine, 'Nature as a Commercial Setting: The Case of Nature-based Tourism Providers in Sweden', *Current Issues in Tourism*, Volume 21, No. 16 (2018), pp. 1893–1911.

Marko, Trogrlić, and Josip, Vrandečić, *French rule in Dalmatia, 1806–1814: globalizing a local geopolitics*, in Ute Planert (ed), *Napoleon's Empire: European Politics in Global Perspective* (Basingstoke and New York, 2016), pp. 264–276.

Marston, John A., *Cambodian pilgrimage groups in India and Sri Lanka*, in Bruntz, Courtney, Schedneck, Brooke (eds), *Buddhist Tourism in Asia* (Honolulu, 2020), pp. 111–126.

Martins, Luciana L., 'A Naturalist's Vision of the Tropics: Charles Darwin and the Brazilian Landscape', *Singapore Journal of Tropical Geography*, Volume 21, No. 1 (March 2000), pp. 19–33.

Martin, Alison E., *"Fresh fields of exploration"': Cultures of scientific knowledge and Ida Pfeiffer's Second Voyage round the world (1856)*, in Martin, Alison E., Missinne, Lut, van Dam, Beatrix (eds), *Travel Writing in Dutch and German, 1790–1830: Modernity, Regionality, Mobility* (London and New York, 2017), pp. 75–94.

Mateu Bellés, Joan F., and Rosselló i Verger, Vicenç M. (eds), *Cuadernos de Geografía de la Universitat de València*, Volume 62 (1997).

Mathieu, Jon, *Divergent perception: deserts and mountains in transition to modernity, seen through Alexander von Humboldt's Views of nature*, in Boscani Leoni, Simona, Baumgartner, Sarah, Knittel, Meike (eds), *Connecting Territories: Exploring People and Nature, 1700–1850* (Leiden and Boston, 2022), pp. 189–209.

Maryks, Robert Aleksander, and Wright, Jonathan (eds), *Jesuit Survival and Restoration. A Global History, 1773–1900* (Leiden and Boston, 2015).

Masuero, Mariarosa, *Guido Gozzano, Libri e lettere* (Florence, 2005).

May, Helen, Kaur, Baljit, and Prochner, Larry, *Empire, Education, and Indigenous Childhoods: Nineteenth-Century Missionary Infant Schools in Three British Colonies* (Farnham and Burlington VT, 2014).

Mazzarello, Paolo, *Spallanzani, Lazzaro*, in *Dizionario Biografico degli Italiani: Volume 93*, https://www.treccani.it/enciclopedia/lazzaro-spallanzani_%28Dizionario-Biografico%29/ (Rome, 2018).

Mazzarello, Paolo, *Costantinopoli 1786: la congiura e la beffa. L'intrigo Spallanzani* (Turin, 2004).

Mazzeo, Tilar J., *Volney, Constantin François de Chasseboeuf, Comte de*, in *Encyclopedia of the Romantic Era 1760–1850* (New York and London, 2004).

McAteer, William, *Rivals in Eden: A History of the French Settlement and British Conquest of the Seychelles Islands (1742–1818)* (Lewes, 1991).

McAteer, William, *Hard Times in Paradise: The History of Seychelles 1827–1919* (Mahé, 2000).

McCracken Lacy, Lisa, *Lady Anne Blunt in the Middle East: Politics, Travel and the Idea of Empire* (London and New York, 2018).

McDonald, Donald, and Hunt, Leslie B., *A History of Platinum and its Allied Metals* (London, 1982).

McElroy, Ann, *Nunavut Generations: Change and Continuity in Canadian Inuit Communities* (Long Grove, 2008).

McNeill, John Robert, 'Observations on the Nature and Culture of Environmental History', *History and Theory* Volume 42, No. 4 (2003), pp. 5–43.

Meeks, Joshua, *France, Britain, and the Struggle for the Revolutionary Western Mediterranean: War, Culture and Society, 1750–1850* (Cham, 2017).

Mellersh, Howard E.L., *FitzRoy of the Beagle* (London, 1968).

Merchant, Carolyn, *The Columbia Guide to American Environmental History* (New York, 2002).

Mexia, Teresa, Vieira, Joana, Príncipe, Adriana, Anjos, Andreia, Silva, Patrícia, Lopes, Nuno, Freitas, Catarina, Santos-Reis, Margarida, Correia, Otília, Branquinho, Cristina, and Pinho, Pedro, *Ecosystem services: urban parks under a magnifying glass'*, *Environmental Research*, Volume 160 (Janaury 2018), pp. 469–478.

Micheletti, Cesare, *Landscape value*, in Dolomites UNESCO World Heritage, https://www.dolomitiunesco.info/universal-values-dolomites-unesco/il-valore-del-paesaggio/?lang=en, accessed on December 2, 2022.

Milani, Raffaele, *The Art of the Landscape* (Montreal and Kingston, 2009).

Miller, Andrew P., *Sustainable Ecotourism in Central America: Comparative Advantage in a Globalized World* (Lanham and London, 2016).

Mills, John Saxon, *The Panama Canal: A History and Description of the Enterprise* (London, 1913).

Mingay, Gordon Edmund, *Young, Arthur*, in *Oxford Dictionary of National Biography*, https://doi.org/10.1093/ref:odnb/30256, accessed on December 2, 2022 (Oxford, 2004).

Ministero della Pubblica Istruzione (ed), *Ruoli di anzianità al 1° agosto 1902, edizione provvisoria* (Rome, 1902).

Moioli, Angelo, *La gelsibachicoltura nella campagne lombarde dal Seicento alla prima metà dell'Ottocento* (Trento, 1981).

Moon, Brenda E., 'Marianne North's *Recollections of a Happy Life*: How They Came to be Written and Published', *Journal of the Society for the Bibliography of Natural History*, Volume 8, No. 4 (1978), pp. 497–505.

Moore, David T., 'Sir William Hamilton's Volcanology and his Involvement in *Campi Phlegræi*', *Archives of Natural History*, Volume 21, No. 2 (1994), pp. 169–193.

Morel, Jean Paul, *Monplaisir, un jardin bien nommé. Deuxième partie : 1767–1772*, http://www.pierre-poivre.fr/Monplaisir-P2.pdf, accessed on October 19, 2022.

Morello, Nicoletta, *Lazzaro Spallanzani geopaleontologo dall'origine delle sorgenti alla vulcanologia*, in Montalenti, Giuseppe, Rossi, Paolo (eds), *Lazzaro Spallanzani e la biologia del Settecento. Teorie, esperimenti, istituzioni scientifiche* (Florence, 1982), pp. 271–281.

Morgan, David, 'Sources of Enlightenment: The Idealizing of China in the Jesuits' *Lettres édifiantes and Voltaire's Siècle de Louis XIV'*, *Romance Notes*, Volume 37, No. 3 (1997), pp. 263–272.

Morgan, Susan, *Introduction*, in Susan Morgan (ed), *North, Marianne Recollections of a Happy Life: Volume I* (Charlottesville and London, 1993), pp. xii–xl.

Mowforth, Martin, and Munt, Ian, *Tourism and Sustainability: New Tourism in the Third World* (London, 1998).

Murawska-Muthesius, Katarzyna, *Imaging and Mapping Eastern Europe: Sarmatia Europea to Post-Communist Bloc* (New York and Abingdon, 2021).

Murphy, Patricia, *In Science's Shadow: Literary Constructions of Late Victorian Women* (Columbia and London, 2006).

Musa, Ghazali, Higham, James, and Thompson-Carr, Anna (eds), *Mountaineering Tourism* (Abingdon and New York, 2015).

Naidoo, Robin, and Adamowicz, Wiktor L., 'Biodiversity and Nature-based Tourism at Forest Reserves in Uganda', *Environment and Development Economics*, Volume 10, No. 2 (May 2005), pp. 159–178.

National Geographic and Oceana (eds), *Islas Desventuradas: biodiversidad marina y propuesta de conservación. Informe de la expedición 'Pristine Seas' National Geographic Society / Oceana, febrero del 2013*, https://media.nationalgeographic.org/assets/file/PristineSeasDesventuradasScientificReport.pdf?_gl=1*vjdwpf*_ga*MTg4NzQzODEyOC4xNjU3NzE4NzEz*_ga_JRRKGYJRKE*MTY1NzcxODcxMy4xLjEuMTY1NzcxODcxOC4w, accessed on October 19, 2022.

National Geospatial-Intelligence Agency, *Nautical Publications*: https://msi.nga.mil/Publications/APN, accessed on December 15, 2022.

Navarrini, Roberto, *Cenni biografici di Giuseppe Acerbi*, in Navarrini, Roberto (ed), *Le Carte Acerbi nella Biblioteca Teresiana di Mantova. Inventario* (Rome, 2002), pp. vii–xiii.

Nazzaro, Antonio, *Il Vesuvio. Storia eruttiva e teorie vulcanologiche* (Naples, 2001).

Negm, Abdelazim M., *The Nile River* (Cham, 2017).

Neumann, Lucien, *Biographies vétérinaires* (Paris, 1896).

Newsome Crossley, John, *The Dasmariñases, Early Governors of the Spanish Philippines* (Abingdon and New York, 2016).

Noltie, Henry J., *The Botanical Collections of Colonel and Mrs Walker: Ceylon, 1830–1838* (Edinburgh, 2013).

Noltie, Henry J., *Robert Wight and his European Botanical Collaborators*, in Damodaran, Vinita, Winterbottom, Anna, Lester, Alan (eds), *The East India Company and the Natural World* (London, 2014), pp. 58–79.

Northcott, Cecil, *Glorious Company: One Hundred and Fifty Years Life and Work of the London Missionary Society 1795–1945* (London, 1945).

Nyberg, Kenneth, *Linnaeus's Apostles and the globalization of knowledge, 1729–1756*, in Manning, Patrick, and Rood, Daniel (eds), *Global Scientific Practice in an Age of Revolutions, 1750–1850* (Pittsburgh, 2016), pp. 73–89.

O'Brien, Anne E., *Lady Morgan's travel writing on Italy: a novel approach*, in Conroy, Jane (ed), *Cross-Cultural Travel: Papers from the Royal Irish Academy Symposium on Literature and Travel* (New York, 2003), pp. 179–185.

Office of International Affairs National Research Council (ed), *Application of Biotechnology to Traditional Fermented Foods: Report of an Ad Hoc Panel of the Board on Science and Technology for International Development* (Washington DC, 1992).

Ohno, Kenichi, *The History of Japanese Economic Development: Origins of Private Dynamism and Policy Competence* (Abingdon and New York, 2017).

O'Rourke, Kevin H., and Williamson, Jeffrey G., *Globalization and History: The Evolution of a Nineteenth-Century Atlantic Economy* (New York, 1999).

Ottaviani, Alessandro, *Olivi, Giuseppe*, in *Dizionario Biografico degli Italiani: Volume 79*, https://www.treccani.it/enciclopedia/giuseppe-olivi_%28Dizionario-Biografico%29/, accessed on December 15, 2022 (Rome, 2013).

Oulebsir, Nabila, *From ruins to heritage: the past perfect and the idealized antiquity in North Africa*, in Klaniczay, Gábor, Werner, Michael, and Gecser, Ottó (eds), *Multiple Antiquities – Multiple Modernities: Ancient Histories in Nineteenth Century European Cultures* (Frankfurt and New York, 2011).

Palladino, Monica, Cafiero, Carlo, and Marcianò, Claudio, *The role of social relations in promoting effective policies to support diversification within a fishing community in Southern Italy*, in Calabrò, Francesco, Della Spina, Lucia, Bevilacqua, Carmelina (eds), *New Metropolitan Perspectives: Local Knowledge and Innovation Dynamics towards Territory Attractiveness through the Implementation of Horizon/E2020/Agenda2030: Volume 2* (Cham, 2019), pp. 124–133.

Pantaleoni, Marco, Console, Fabiana, Lorusso, Lorenzo, Petti, Fabio Massimo, Franchini, Antonia Francesca, Porro, Alessandro, and Romano, Marco, *Italian physicians' contribution to geosciences*, in Duffin, Christopher J., Gardner-Thorpe, Christopher, Moody, Richard T. J. (eds), *Geology and Medicine: Historical Connections* (London, 2017), pp. 55–75.

Paschoud, Adrien, *Le Monde amérindien au miroir des Lettres édifiantes et curieuses* (Oxford, 2000).

Pasta, Renato, *Santi, Giorgio*, in *Dizionario Biografico degli Italiani: Volume 90* (Rome, 2017).

Patterson, Sheila, *Tasty little dishes of the cape*, in Kuper, Jessica (ed), *The Anthropologist's Cookbook* (Abingdon and New York, 1997, Revised edition), pp. 114–121.

Payne, Michelle, *Marianne North: A Very Intrepid Painter* (Chicago and Kew, 2016, 2nd edition).

Pazzagli, Rossano, *Il sapere dell'agricoltura. Istruzione, cultura, economia nell'Italia dell'Ottocento* (Milan, 2008).

Pederzani, Ivana, *I Dandolo. Dall'Italia dei Lumi al Risorgimento* (Milan, 2014).

Perkins, Adam J., and Dick, Steven J., *The British and American Nautical Almanacs in the 19th Century*, in Seidelmann, Kenneth P., Hohenkerk, Catherine Y. (eds), *The History of Celestial Navigation: Rise of the Royal Observatory and Nautical Almanacs* (Cham, 2020), pp. 157–197.

Peters, Joe, 'Sharing National Park Entrance Fees: Forging New Partnerships in Madagascar', *Society and Natural Resources*, Volume 11, No. 5 (1998), pp. 517–530.

Pipino, Giuseppe, 'L'oro del Monte Rosa e la sua storia', *Bollettino Storico per la Provincia di Novara*, Volume 91, No. 2 (2000), pp. 321–352.

Pizzamiglio, Gilberto, *Introduzione*, in Alberto Fortis, *Viaggio in Dalmazia*, Eva Viani (ed) (Venice, 1987), pp. ix–xxx.

Pomeroy, Sarah B., and Kathirithamby, Jeyaraney, *Maria Sibylla Merian: Artist, Scientist, Adventurer* (Los Angeles, 2018).

Puccinelli, Elena, *Greppi, Antonio*, in *Dizionario Biografico degli Italiani: Volume 59*, https://www.treccani.it/enciclopedia/antonio-greppi_%28Dizionario-Biografico%29/ accessed on December 4, 2022 (Rome, 2002a).

Puccinelli, Elena, *Greppi, Paolo*, in *Dizionario Biografico degli Italiani: Volume 59* https://www.treccani.it/enciclopedia/paolo-greppi_%28Dizionario-Biografico %29/ accessed on December 4, 2022 (Rome, 2002b).

Pulteney, Richard, *A General View of the Writings of Linnaeus* (London, 1805, 2nd edition).

Radkau, Joachim. *Nature and Power: A Global History of the Environment*, Dunlap, Thomas (trans), (New York, 2008).

Räikkönen, Juulia, Grénman, Miia, Rouhiainen, Henna, Honkanen, Antti, and Sääksjärvi, Ilari E, 'Conceptualizing Nature-based Science Tourism: A Case Study of Seili Island, Finland', *Journal of Sustainable Tourism* (2021), https://doi.org/ 10.1080/09669582.2021.1948553, accessed on December 22, 2022.

Ratcliffe, Derek, *Lapland: A Natural History*, illustrated by Mike Unwin (London, 2005).

Raza, Rosemary, *In Their Own Words: British Women Writers and India 1740–1857* (Oxford, 2006).

Real Sociedad Económica de Amigos del País de Valencia (ed), *Antonio José Cavanilles (1745–1804): Segundo centenario de la muerte de un gran botánico* (Valencia, 2004).

Reed, Marcia, *A perfume is best from afar: publishing China for Europe*, in Reed, Marcia, Demattè, Paola (eds), *China on Paper: European and Chinese Works from the Late Sixteenth to the Early Nineteenth Century* (Los Angeles, 2007), pp. 9–27.

Reed, Michael Charles, and Barnes, James Franklin (eds), *Culture, Ecology, and Politics in Gabon's Rainforest* (Lewiston, 2003).

Reinert, Sophus A., *The Academy of Fisticuffs: Political Economy and Commercial Society in Enlightenment Italy* (Cambridge MA and London, 2018).

Restif-Filliozat, Manonmani, 'The Jesuit Contribution to the Geographical Knowledge of India in the Eighteenth Century', *Journal of Jesuit Studies*, Volume 6, No. 1 (2019), pp. 71–84.

Ritchie, J.R.Brent, and Crouch, Geoffrey I., *The Competitive Destination: A Sustainable Tourism Perspective* (Wallingford and Cambridge MA, 2003).

Ritchie, Leitch, and Stanfield, Clarkson, *Travelling Sketches in the North of Italy, the Tyrol, and on the Rhine* (London, 1832).

Richards, John F. *The Unending Frontier: An Environmental History of the Early Modern World* (Berkeley, Los Angeles, and London, 2003).

Riva, Elena, *Da negoziante a gentiluomo. La formazione di Paolo Greppi tra commercio, finanza e diplomazia*, in Mafrici, Mirella (ed), *Rapporti diplomatici e scambi commerciali nel Mediterraneo moderno* (Salerno and Soveria Mannelli, 2004), pp. 379–444.

Rivers, W.H.R., 'Irrigation and the Cultivation of Taro', *Memoirs and Proceedings of the Manchester Literary and Philosophical Society*, Volume 60 (1915–1916), pp. xliv–xlv.

Robinson-Tomsett, Emma, *Women, Travel and Identity: Journeys by Rail and Sea, 1870–1940* (Manchester, 2013).

Robinson, Michael F., *Scientific travel*, in Das, Nandini, Youngs, Tim (eds), *The Cambridge History of Travel Writing* (Cambridge and New York, 2019), pp. 488–503.

Rocca, Giuseppe, *Dal prototurismo al turismo globale. Momenti, percorsi di ricerca, casi di studio* (Turin, 2013).

Roe, Alan D., *Into Russian Nature: Tourism, Environmental Protection, and National Parks in the Twentieth Century* (Oxford and New York, 2020).

Role, André, *Un destin hors série: la vie aventureuse d'un savant: Bory de Saint-Vincent 1778–1846* (Paris, 1973).

Roman, John F. (ed), *Medieval Travel and Travellers: A Reader* (Toronto, Buffalo and London, 2020).

Romani, Marzio Achille, *'Alpe' e 'Alpi'. Economie e società della montagna tra Medioevo e XIX secolo* (Brescia, 1987).

Romano, Maurizio, *Alle origini dell'industria lombarda. Manifatture, tecnologie e cultura economica nell'età della Restaurazione* (Milan, 2012).

Ross-Bryant, Lynn, *Pilgrimage to the National Parks: Religion and Nature in the United States* (Abingdon and New York, 2013).

Ross, Corey, *Ecology and Power in the Age of Empire: Europe and the Transformation of the Tropical World* (Oxford, 2017).

Ross, Robert, *A Concise History of South Africa* (Cambridge and New York, 2nd edition).

Roßbach, Nikola (ed), *Der See ging hoch mit seinen blauen, blauen, ach, so reizend blauen Wellen. Literatur zum Gardasee aus drei Jahrhunderten* (Vienna, 2014).

Roth, Ralf, and Dinhobl, Günter (eds), *Across the Borders: Financing the World's Railways in the Nineteenth and Twentieth Centuries* (Aldershot and Burlington VT, 2008).

Routledge, Karen, *Do You See Ice? Inuit and American at Home and Away* (Chicago and London, 2018).

Ruffolo, Ida, *The Perception of Nature in Travel Promotion Texts: A Corpus-based Discourse Analysis* (Bern, 2015).

Rupke, Nicolaas A., *Alexander von Humboldt: a Metabiography* (Frankfurt am Main, 2005).

Sachs, Jeffrey D., *The Ages of Globalization: Geography, Technology, and Institutions* (New York, 2020).

Samarasinghe, Dinal J.S., *The Human-Crocodile Conflict in Nilwala River, Matara (Phase 1)* (Colombo, 2014).

Sandri, Giulio, *Elogio del dottor Ciro Pollini* (Verona, 1833).

Saunders, Kay (ed), *Indentured Labor in the British Empire, 1834–1920* (London and Canberra, 1983).

Saxena, Gunjan, 'Clark, Gordon, Oliver, Tove, and Ilbery, Brian, 'Conceptualizing Integrated Rural Tourism'', *Tourism Geographies*, Volume 9, No. 4 (2007), pp. 347–370.

Scarr, Deryck, *Seychelles since 1770: History of a Slave and Post-Slavery Society* (London, 2000).

Schiebinger, Londa, and Swan, Claudia, *Colonial Botany: Science, Commerce, and Politics in the Early Modern World* (Philadelphia, 2005).

Schmidt, Benjamin, *Inventing Exoticism: Geography, Globalism, and Europe's Early Modern World* (Philadelphia, 2015).

Scirocco, Alfonso, *Colletta, Pietro*, in *Dizionario Biografico degli Italiani: Volume 27*, https://www.treccani.it/enciclopedia/pietro-colletta_%28Dizionario-Biografico %29/, accessed on December 16, 2022 (Rome, 1982).

Seppälä, Matti (ed), *The Physical Geography of Fennoscandia* (Oxford and New York, 2005).

Seymour, Miranda, *Mary Shelley* (New York, 2000).

Shacklefold, Peter, *A History of the World Tourism Organization* (Bingley, 2020).

Short Safaris and Tracks, https://gabon-adventuretours.ga, accessed on November 14, 2022.

Sidali, Katia Laura, Spiller, Achim, and Schulze, Birgit (eds), *Food, Agri-Culture and Tourism: Linking Local Gastronomy and Rural Tourism: Interdisciplinary Perspectives* (Berlin and Heidelberg, 2011).

Sigrist, René (ed), *H.-B. de Saussure (1740–1799): un regard sur la Terre* (Geneva, 2001).

Sigrist, René, *La Nature à l'épreuve. Les débuts de l'expérimentation à Genève (1670–1790)* (Paris, 2011).

Simões, Ana, Carneiro, Ana, and Diogo, Maria Paula (eds), *Travels of Learning: A Geography of Science in Europe* (Dordrecht, Boston and London, 2003).

Simonton, Deborah, *A History of European Women's Work: 1700 to the Present* (London and New York, 1998).

Skjöldebrand, Anders Friedrik, *Voyage pittoresque au Cap Nord* (Stockholm 1801–1802).

Slee, Bill, *Rural tourism and recreation in the EU*, in Davidova, Sophia M., Thomson, Kenneth J., Mishra, Ashok K. (eds), *Rural Policies and Employment: TransAtlantic Experiences* (Singapore, 2019), pp. 333–350.

Sloan, Philip, Legrand, Willy, and Kinski, Sonja, 'The Restorative Power of Forests: The Tree House Hotel Phenomena in Germany', *Advances in Hospitality and Leisure*, Volume 12 (2016), pp. 181–189.

Slocum, Susan L., Kline, Carol, and Holden, Andrew (eds), *Scientific Tourism: Researchers and Travellers* (Abingdon and New York, 2015).

Smethurst, Paul, *Travel Writing and the Natural World, 1748–1840* (Basingstoke and New York, 2012).

Smith, Melanie, and Richards, Greg (eds), *The Routledge Handbook of Cultural Tourism* (Abingdon and New York, 2013).

Sörbom, Per, *Diderot's Russian University*, in Göranzon, Bo, Florin, Magnus (eds), *Skill and Education: Reflection and Experience* (London, 1992), pp. 195–206.

Soto Arango, Diana, E., 'Cavanilles y Zea: una amistad político-científica', *Asclepio: Revista de historia de la medicina y de la ciencia*, Volume 47, No. 1 (1995), pp. 169–196.

Southey, Robert, *History of Brazil: Volume III* (London, 1819).

Spary, Emma C., *Utopia's Garden: French Natural History from Old Regime to Revolution* (Chicago and London, 2000).

Spary, Emma C., *Eating the Enlightenment: Food and the Sciences in Paris* (Chicago and London, 2012).

Sprang, Felix, *Charles Darwin, The Voyage of the Beagle (1839)*, in Shaff, Barbara (ed), *Handbook of British Travel Writing* (Berlin and Boston, 2020), pp. 373–395.

Stafford, Paul, *The Best Beagle Channel Tours & Cruises from Ushuaia*, August 5, 2020: https://www.travelmag.com/articles/beagle-channel-tours/, accessed on November 15, 2022.

Stagl, Justin, *A History of Curiosity: The Theory of Travel 1550–1800* (Abingdon, 2005, 2nd edition).

Stanley, Brian (ed), *Christian Missions and the Enlightenment* (Abingdon and New York, 2001).

Stiefel, Barry L., *Maple: the sugar of abolitionist aspirations*, in Victoria, Barnett-Woods (ed), *Cultural Economies of the Atlantic World: Objects and Capital in the Transatlantic Imagination* (New York and Abingdon, 2020), pp. 147–172.

Stokes, George T. (ed), *Pococke's Tour in Ireland in 1752* (Dublin and London, 1891).

Stone, Lesego Senyana, 'Perceptions of Nature-based Tourism, Travel Preferences, Promotions and Disparity between Domestic and International Tourists: The Case of Botswana' (Arizona State University, 2014), https://keep.lib.asu.edu/_flysystem/fedora/c7/124454/Stone_asu_0010E_14487.pdf, accessed on October 21, 2022.

Stronza, Amanda L., Hunt, Carter A., and Fitzgerald, Lee A., 'Ecotourism for Conservation?', *Annual Review of Environment and Resources*, Volume 44 (2019), pp. 229–253.

Sunstein, Emily W., *Mary Shelley: Romance and Reality* (Baltimore, 1989).

Surić, Maša, Lončarić, Robert, Čuka, Anica, and Faričić, Josip, 'Geological issues in Alberto Fortis' *Viaggio in Dalmazia* (1774) – Excursions géologiques extraites du *Viaggio in Dalmazia* d'Alberto Fortis', *Comptes Rendus Geoscience*, Volume 339, No. 9 (August 2007), pp. 640–650.

Sweet, Rosemary, *Cities and the Grand Tour: The British in Italy, c. 1690–1820* (Cambridge and London, 2012).

Tagarelli, Giuseppe, and Torchia, Francesco (eds), *Turismo, paesaggio e beni culturali. Prospettive di tutela, valorizzazione e sviluppo sostenibile* (Rome, 2021).

Tapia-Guerra, Jan M., Mecho, Ariadna, Easton, Erin E., de los Ángeles Gallardo, María, Gorny, Matthias, and Sellanes, Javier, 'First Description of Deep Benthic Habitats and Communities of Oceanic Islands and Seamounts of the Nazca Desventuradas Marine Park Chile', *Scientific Reports*, Volume 11, No. 6209 (2021), https://doi.org/10.1038/s41598-021-85516-8, accessed on October 19, 2022.

Taylor, James, *The Voyage of the Beagle: Darwin's Extraordinary Adventure aboard FitzRoy's Famous Survey Ship* (London, 2008).

Taylor, Kenneth L., *Dolomieu, Dieudonné (called Déodat) de Gratet de*, in Gillispie, Charles (ed), *Dictionary of Scientific Biography: Volume 4* (New York, 1971), pp. 149–153.

Taylor, Steve, Varley, Peter, and Johnston, Tony (eds), *Adventure Tourism: Meanings, Experience, and Learning* (Abingdon and New York, 2013).

Tempesta, Tiziano, *La valutazione del paesaggio*, in Marangon, Francesco (ed), *Gli interventi paesaggistico-ambientali nelle politiche regionali di sviluppo rurale* (Milan, 2006), pp. 58–76.

The Journal of Modern History, Volume 93, No. 1: *Travelers* (March 2021).

Thomson, Emma, *Life on the longest river: sailing 140 miles along the Nile*, August 5, 2019, https://www.nationalgeographic.co.uk/travel/2019/07/egypt-life-longest-river, accessed on November 14, 2022.

Tilley, Helen, *Africa as a Living Laboratory. Empire, Development, and the Problem of Scientific Knowledge, 1870–1950* (Chicago and London, 2011).

Tisdell, Clem, and Wilson, Clevo, *Nature-Based Tourism and Conservation: New Economic Insights and Case Studies* (Cheltenham UK and Northampton MA, 2012).

Tornedo, Recaredo S., *Chile Ilustrado: Guía descriptiva del territorio de Chile, de las capitales de provincia, i de los puertos principales* (Valparaiso, 1872).

Torriani, Edoardo (ed), 'Alcuni documenti relativi ad Emanuele Haller, in relazione al suo palazzo di Mendrisio (1794–1818)', *Bollettino Storico della Svizzera Italiana*, Volume 18 (1896), pp. 19–24.

Troelstra, Anne S., *Bibliography of Natural History Travel Narratives* (Leiden, 2016).

Trusted, Jennifer, *Beliefs and Biology: Theories of Life and Living* (Basingstoke and New York, 2003, 2nd edition).

Turner, Lance C., 'Botanizing in the Borderlands: the Limits of Scientific Indigeneity in Late Colonial New Spain', *Colonial Latin American Review*, Volume 30, No. 1 (2021), pp. 109–136.

Turri, Eugenio, *Il paesaggio come teatro. Dal territorio vissuto al territorio come rappresentato* (Venice, 1998).

Tyler, Duncan, and Dangerfield, J. Mark, 'Ecosystem Tourism: A Resource-based Philosophy for Ecotourism', *Journal of Sustainable Tourism*, Volume 7, No. 2 (1999), pp. 146–158.

Tzanelli, Rodanthi, *Thanatourism and Cinematic Representations of Risk: Screening the End of Tourism* (Abingdon and New York, 2016).

Ünal, Vahdet, Ertör, Irmak, Ertör-Akyazi, Pinar, and Tunca, Sezgin, *Making pescatourism just for small-scale fisheries: the case of Turkey and lessons for others*, in Jentoft, Svein, Chuenpagdee, Ratana, Bugeja Said, Alicia, Isaacs, Moenieba (eds), *Blue Justice: Small-Scale Fisheries in a Sustainable Ocean Economy* (Cham, 2022), pp. 315–333.

UNESCO-World Heritage Convention (n.d.): http://whc.unesco.org/en/list/1133

UNESCO-World Heritage Germany (n.d.): https://worldheritagegermany.com/germanys-ancient-beech-forests/

Unger, Richard W., 'Dutch Ship Design in the Fifteenth and Sixteenth Centuries', *Viator: Medieval and Renaissance Studies*, Volume 4 (1973), pp. 387–411.

UNWTO-Tourism in the 2030 Agenda, https://www.unwto.org/tourism-in-2030-agenda, accessed on December 22, 2022.

Upton, William Henry, *Upton Family Records: Being Genealogical Collections for an Upton Family History* (London, 1893).

Vai, Gian Battista, '*Light and shadow: the status of Italian geology around 1807*, in Lewis, Cherry L. E., Knell, Simon J. (eds), *The Making of the Geological Society of London* (London, 2009), pp. 179–202.

Valdés, Benito, 'Early Botanical Exploration of the Maghreb', *Flora Mediterranea*, Volume 31 (2021), pp. 5–18.

Välimaa, Jussi, *A History of Finnish Higher Education from the Middle Ages to the 21st Century* (Cham, 2019).

Vander Naald, Brian, 'Examining Tourist Preferences to Slow Glacier Loss: Evidence from Alaska', *Tourism Recreation Research*, Volume 45, No. 1 (2020), pp. 107–117.

Väre, Henry, 'Fredrik Wilhelm Radloff – Demonstrator in Botany at old Åbo Akademi', *Memoranda Societatis Pro Fauna Flora Fennica*, Volume 92 (2016), pp. 92–98.

Väre, Henry, and Ulvinen, Tauno, 'J. Julinin, K. H. Eberhardtin ja H. S. Zidbäckin julkaisemattomia kasvitietoja 1800-luvulta etenkin Oulusta ja muualta Pohjois-Suomesta', *Norrlinia*, Volume 12 (2005), pp. 1–58.

Vermeulen, Han F., *Before Boas: the Genesis of Ethnography and Ethnology in the German Enlightenment* (Lincoln NE, 2015).

Vialette, Yannick, Mao, Pascal, and Bourlon, Fabien, 'Scientific Tourism in the French Alps: A Laboratory for Scientific Mediation and Research', *Journal of Alpine Research – Revue de géographie alpine*, Volume 109, No. 2 (2021), https://doi.org/10.4000/rga.9189, accessed on December 2, 2022.

Vicuña Urrutia, Manuel, *Un juez en los infiernos: Benjamín Vicuña Mackenna* (Santiago de Chile, 2009).

Vieira, Joana, Matos, Paula, Mexia, Teresa, Silva, Patrícia, Lopes, Nuno, Freitas, Catarina, Correia, Otília, Santos-Reis, Margarida, Branquinho, Cristina, and Pinho, Pedro, 'Green Spaces Are Not All the Same for the Provision of Air Purification and Climate Regulation Services: The Case of Urban Parks', *Environmental Research*, Volume 160, 306–313 (Janaury 2018).

Visconti, Agnese, *Scienziati e naturalisti dai Balcani a Capo Nord*, in *Europa: Storie di viaggiatori italiani* (Milan, 1988), pp. 200–229.

Visconti, Agnese, *Paesaggi di Lombardia: il caso dell'ulivo tra ambienti naturali e tecniche manifatturiere (1772–1796)*, in Guerci, Gabriella, Pelissetti, Laura, and Scazzosi, Lionella (eds), *Oltre il giardino: le architetture vegetali e il paesaggio* (Florence, 2003), pp. 167–174.

Visconti, Agnese, 'Carlo Amoretti in viaggio tra Lombardia Austriaca e Mendrisiotto (1791): sentimenti d'amore e interessi scientifici', *Archivio Storico Ticinese*, No. 157 (2015), pp. 108–123.

Vodopivec, Peter, *Illyrian provinces from a Slovene perspective: myth and reality*, in Planert, Ute (ed), *Napoleon's Empire: European Politics in Global, Napoleon's Empire: European Politics in Global Perspective Perspective* (Basingstoke and New York, 2016), respectively pp. 252–263 (Basingstoke and New York, 2016).

Volga Dream – Russian River Cruises: Moscow to Astrakhan 14 days-13 nights, https://www.volgadream.com/cruises/moscow-to-astrakhan/, accessed on November 14, 2022.

Wacker, Peter O., *Swedish settlement in New Jersey before 1800*, in Hoffecker, Carol E., Waldron, Richard, Williams, Lorraine E., Benson, Barbara E. (eds), *New Sweden in America* (Newark and London, 1995).

Wanner, Kurt, *Lo Spluga. Il passo sublime* (Chiavenna, 2005).

Warde, Paul, *Ecology, Economy and State Formation in Early Modern Germany* (Cambridge and New York, 2006).

Warde, Paul, *The Invention of Sustainability: Nature and Destiny, c. 1500–1870* (Cambridge and New York, 2018).

Warde, Paul, Robin, Libby, and Sörlin, Sverker (eds), *The Environment: A History of the Idea* (Baltimore, 2018).

Weaver, David, 'The Distinctive Dynamics of Exurban Tourism', *International Journal of Tourism Research*, Volume 7 (2005), pp. 23–33.

West, Paige, 'Tourism and Science and Science as Tourism: Environment, Society, Self, and Other in Papua New Guinea', *Current Anthropology*, Volume 49, No. 4 (August 2008), pp. 597–626.

Whelan, Tensie (ed), *Nature Tourism: Managing for the Environment* (Washington D.C., 1991).

Wilkie, David S., and Carpenter, Julia, 'Can nature tourism help finance protected areas in the Congo Basin?', *Oryx*, Volume 33, No. 4 (October 1999), pp. 22–31.

Willis, Sam, *Fighting at Sea in the Eighteenth Century: The Art of Sailing Warfare* (Woodbridge, 2008).

Wilson, Eric G., *The Spiritual History of Ice: Romanticism, Science, and the Imagination* (New York and Basingstoke, 2003).

Wolff, Larry, *Venice and the Slaves: the Discovery of Dalmatia in the Age of Enlightenment* (Stanford, 2001).

Worden, Nigel, *Adjusting the emancipation: freed slaves and farmers in the mid-nineteenth-century South-Western Cape*, in James, Wilmot G., Simons, Mary (eds), *Class, Caste and Color: A Social and Economic History of the South African Western Cape* (New Brunswick NJ, 1992), pp. 31–39.

World Tourism Organization, *Glossary of Tourism Terms*: https://www.unwto.org/glossary-tourism-terms, accessed on October 17, 2022.

Wormbs, Nina (ed), *Competing Arctic Futures: Historical and Contemporary Perspectives* (Cham, 2018).

Worster, Donald (ed), *The Ends of the Earth: Perspectives on Modern Environmental History* (Cambridge and New York, 1988).

Wulf, Andrea, *The Invention of Nature: the Adventures of Alexander von Humboldt, the Lost Hero of Science* (New York, 2015).

Zemon Davis, Natalie, *Women on the Margins: Three Seventeenth-Century Lives* (Cambridge MA, 1997).

Zanini, Andrea, *Un secolo di turismo in Liguria. Dinamiche, percorsi, attori* (Milan, 2012).

Zanon, Bruno (ed), *Le Dolomiti. Patrimonio mondiale UNESCO. Fenomeni geologici e paesaggi umani* (Pisa, 2021).

Zhang, Chunyan, Knight, David W., Li, Yajuan, Zhou, Yi, Zhou, Meng, and Zi, Minggui, 'Rural Tourism and Evolving Identities of Chinese Communities in Forested Areas', *Journal of Sustainable Tourism*, https://doi.org/10.1080/09669582. 2022.2155829, accessed on December 22, 2022, (2022).

Zhesheng, Ouyang, 'The "Beijing Experience" of Eighteenth-century French Jesuits: A Discussion Centered on *Lettres édifiantes et curieuses écrites des missions étrangères*', *Chinese Studies in History*, Volume 46, No. 2 (2012), pp. 35–57

Zinkina, Julia, Christian, David, Grinin, Leonid, Ilyin, Ilya, Andreev, Alexey, Aleshkovski, Ivan, Shulgin, Sergey, Korotayev, Andrey (eds), *A Big History of Globalization: The Emergence of a Global World System* (Cham, 2019).

Zuelow, Eric G.E., *A History of Modern Tourism* (London and New York, 2016).

Zuelow, Eric G.E., and James, Kevin J. (eds), *The Oxford Handbook of the History of Tourism and Travel* (Oxford, 2022).

Index

Pages in *italics* refer figures and pages in **bold** refer tables.